BEFORE THE DELUGE

BEFORE THE DELUGE

The Vanishing World of the Yangtze's Three Gorges

DEIRDRE CHETHAM

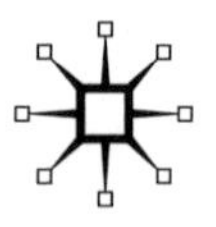

BEFORE THE DELUGE

First published 2002 by
PALGRAVE MACMILLAN™
175 Fifth Avenue, New York, N.Y. 10010 and
Houndmills, Basingstoke, Hampshire, England RG21 6XS.
Companies and representatives throughout the world.

PALGRAVE MACMILLAN is the global academic imprint of the Palgrave Macmillan division of St. Martin's Press, LLC and of Palgrave Macmillan Ltd. Macmillan® is a registered trademark in the United States, United Kingdom and other countries. Palgrave is a registered trademark in the European Union and other countries.

ISBN 0–312–21417–0 hardback

Library of Congress Cataloging-in-Publication Data
Chetham, Deirdre, 1954-
Before the deluge : the vanishing world of the Yangtze's Three Gorges / Deirdre Chetham.
p. cm.
Includes bibliographical references and index.
ISBN 0–312–21417–0
1. Yangtze River Region (China)—History. 2. Yangtze River Region (China)—Social life and customs. 3. Dams—Environmental aspects—China—Yangtze River Gorges. I. Title.

DS793.Y3 C495 2002
951'.38—dc21

2002016939

A catalogue record for this book is available from the British Library.

Design by Letra Libre, Inc.

First edition: October 2002
10 9 8 7 6 5 4 3 2 1

Printed in the United States of America

Sadness of the Gorges

Above the gorges, one thread of sky:
Cascades in the gorges twine a thousand cords.
High up, the slant of splintered sunlight, moonlight:
Beneath, curbs to the wild heave of the waves.
The shock of a gleam, and then another,
In depths of shadow frozen for centuries:
The rays between the gorges do not halt at noon;
Where the straits are perilous, more hungry spittle.
Trees lock their roots in rotted coffins
And the twisted skeletons hang tilted upright:
Branches weep as the frost perches
Mournful cadences, remote and clear.
A spurned exile's shrivelled guts
Scald and seethe in the water and fire he walks through.
A lifetime's like a fine-spun thread,
The road goes up by the rope at the edge.
When he pours his libation of tears to the ghosts in the stream
The ghosts gather, a shimmer on the waves.

—Meng Jiao (751–814)
translated by A.G. Graham

CONTENTS

List of Illustrations

Chapter 1

Chapter 2

Chapter 3

Chapter 4

Chapter 9

CHAPTER 10

CHAPTER 11

CHAPTER 12

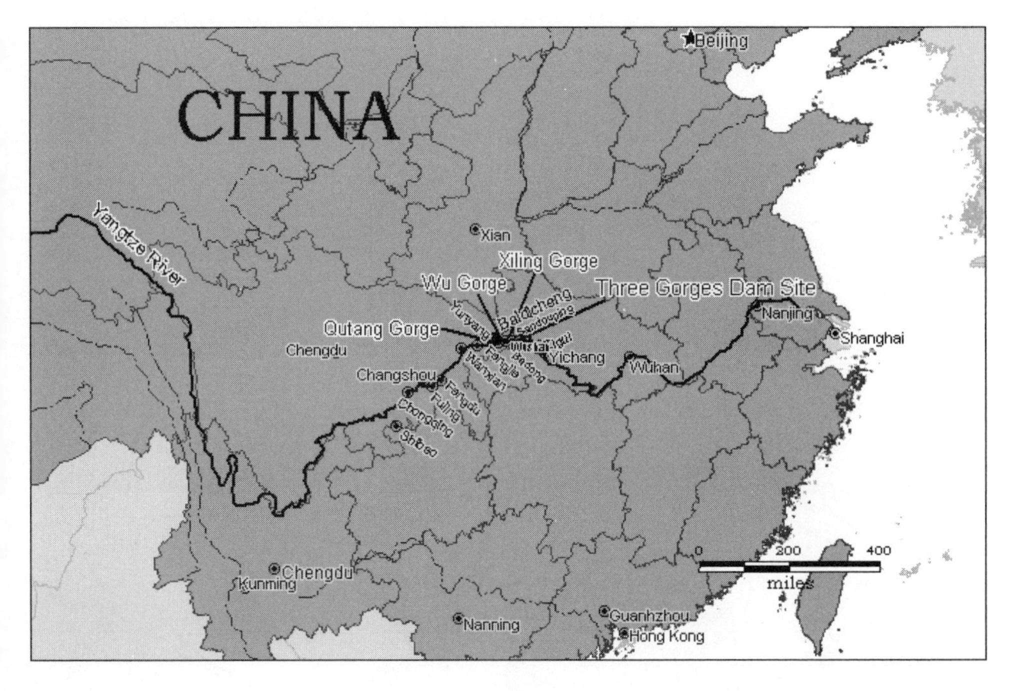

CHINA
Beijing
Yangtze River
Xian
Xiling Gorge
Wu Gorge
Three Gorges Dam Site
Qutang Gorge
Yunyang
Sandouping
Chengdu
Yichang
Changshou
Fengjie
Wanxian
Badong
Fengdu
Fuling
Chongqing
Nanjing
Shanghai
Wuhan
0
200
400
miles
Chengdu
Kunming
Nanning
Guanhzhou
Hong Kong

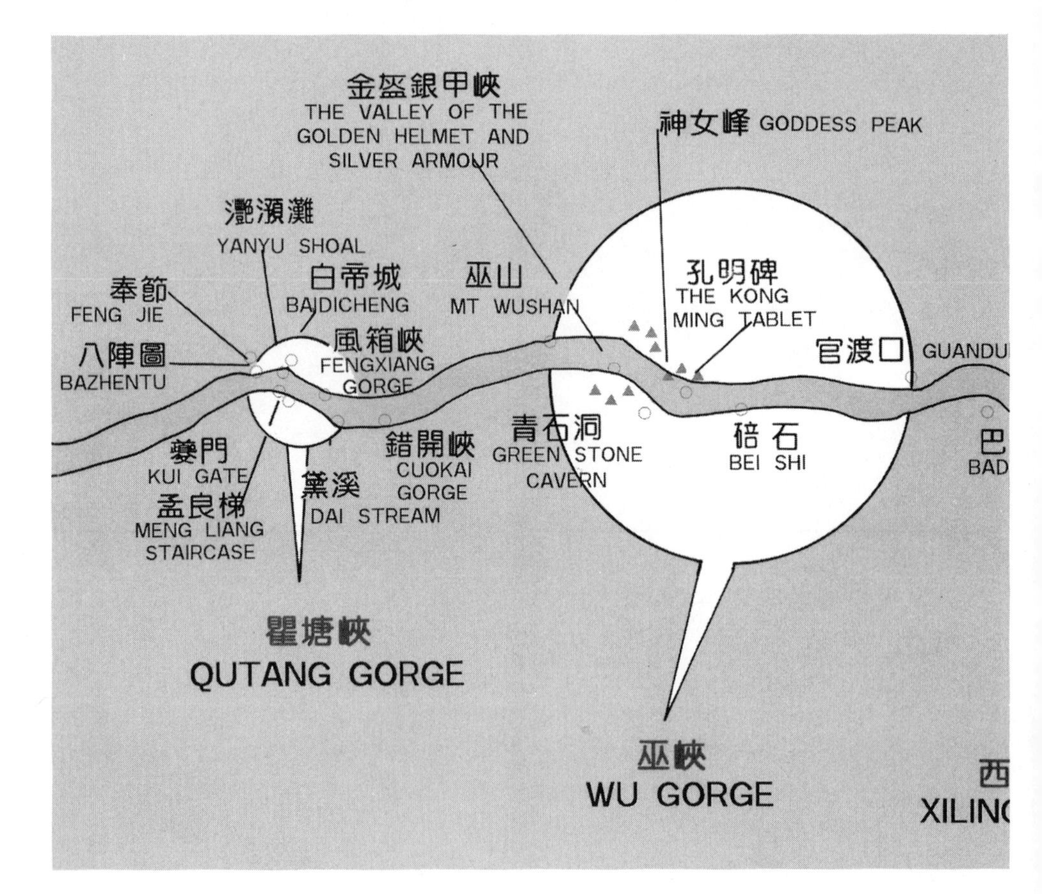

金盔銀甲峽
THE VALLEY OF THE GOLDEN HELMET AND SILVER ARMOUR
神女峰 GODDESS PEAK
灩瀩灘
YANYU SHOAL
奉節
FENG JIE
白帝城
BAIDICHENG
巫山
MT WUSHAN
孔明碑
THE KONG MING TABLET
八陣圖
BAZHENTU
風箱峽
FENGXIANG GORGE
官渡口
GUANDU
青石洞
GREEN STONE CAVERN
碚石
BEI SHI
夔門
KUI GATE
錯開峽
CUOKAI GORGE
巴
BAD
孟良梯
MENG LIANG STAIRCASE
黛溪
DAI STREAM
瞿塘峽
QUTANG GORGE
巫峽
WU GORGE
西
XILIN

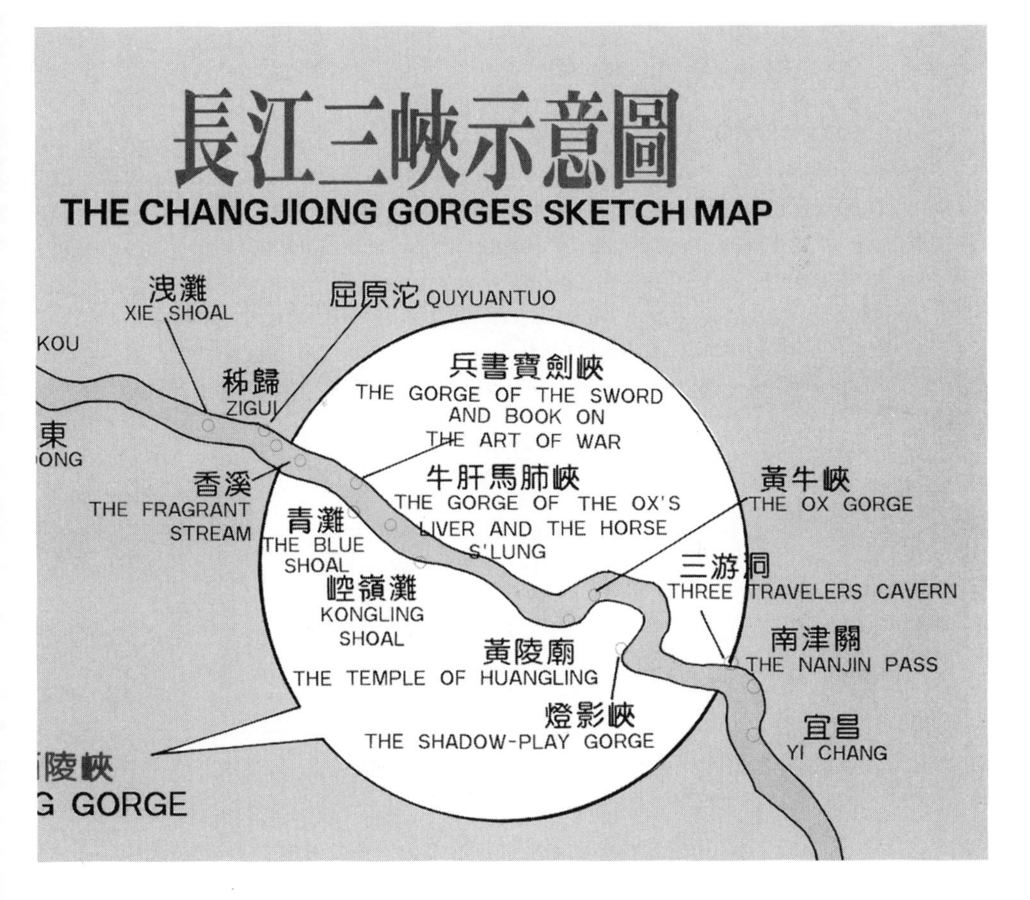
長江三峽示意圖
THE CHANGJIQNG GORGES SKETCH MAP
洩灘
XIE SHOAL
屈原沱 QUYUANTUO
KOU
秭歸
ZIGUI
ONG
香溪
THE FRAGRANT STREAM
兵書寶劍峽
THE GORGE OF THE SWORD AND BOOK ON THE ART OF WAR
牛肝馬肺峽
THE GORGE OF THE OX'S LIVER AND THE HORSE S'LUNG
青灘
THE BLUE SHOAL
崆嶺灘
KONGLING SHOAL
黃陵廟
THE TEMPLE OF HUANGLING
燈影峽
THE SHADOW-PLAY GORGE
黃牛峽
THE OX GORGE
三游洞
THREE TRAVELERS CAVERN
南津關
THE NANJIN PASS
宜昌
YI CHANG
陵峽
G GORGE

CREDITS

THE AUTHOR AND PUBLISHER WISH TO THANK the following authors and publishers for permission to reproduce copyright material:

Penguin Books Ltd. for the poem "Sadness of the Gorges" from *Poems of the Late T'ang,* translated by A. G. Graham (Penguin Classics 1965), copyright A. G. Graham, 1965, reproduced by permission of Penguin Books, Ltd.

Map of Yangtze River Gorges courtesy China International Travel service, Los Angeles, CA. Reprinted by permission.

Map of China by East View Cartographic. Printed by permission.

A Note on Romanization and Measurements

THERE ARE SEVERAL DIFFERENT SYSTEMS of romanizing Chinese characters, the most common still in use being the *pinyin* system, developed in the People's Republic of China and the nineteenth-century Wade-Giles system. I have, for the most part, used the *pinyin* romanization of Chinese words and names but have retained the original spelling in quotes from books and articles. I have also used the original spellings of personal names by which historical figures are familiarly known (for example, Chiang Kai-shek rather than Jiang Jieshi), or according to the preferred spellings of the individuals themselves. A glossary of the most common place-names mentioned in this book follows.

A list of the modern *pinyin* spelling of place-names that occur in this book appears below, along with the previously used Wade-Giles spelling or other spellings and alternate names. Words referring to bodies of water occur frequently in place-names. *Jiang* and *he* both mean river. *Chuan* is an ancient word for river, as in Sichuan, four rivers. *Hu* means lake, as in Hubei and Hunan, north and south of the lake. *Xia* means gorge (thus, *sanxia* is three gorges) and *tan* is rapids, as in *xintan*, or new rapids.

MODERN PINYIN NAMES	OLD SPELLINGS, OTHER NAMES
Badong	Pa-tung
Baidicheng	White King City
Beijing	Peking
Chengdu	Chengtu, Cheng-tu
Chongqing	Chungking, Chung-king
Fengdu	Feng-tu, Fengtu
Fengjie	Kwei Fu, Kwei-chou, Kuizhou
Guangdong	Kwangtung
Guangzhou	Canton

Hangzhou	Hang-chow, Hangchow
Hubei	Hupeh, Hu-pei, Hopeh
Jiangsu	Kiangsu, Kiang-su, Chiang-su
Nanjing	Nanking, Jinling, Chinling
Ningbo	Ningpo
Shaanxi	Shensi
Shandong	Shantung
Shibao Block	Shi-pao-chai
Sichuan	Szechuan, Szecheun, Sze Chuan, Szechwan
Tianjin	Tientsin
Wanxian, Wanzhou	Wan Hsien, Wanhsien
Wuhan, made up of three cities:	
(1) Hankou	Hankow
(2) Hanyang	Hanyang
(3) Wuchang	Wu-ch'ang
Yantai	Chefoo
Yangzhou	Yang-chow, Yangchow
Yibin	Suifu
Yichang	Ichang, I'chang
Zhenjiang	Chin-kiang, Chinkiang, Jingkou
Zhongxian	Chung-chou
Zigui	Kwei, Tse-kwei

The most common Chinese measurements are the *li*, a unit of distance; the *mu*, approximately one-sixth of an acre, used to measure land; and the *jin*, or catty. For the ease of the reader, these have been converted to distances, areas, and weights in the English system. A table of conversion of Chinese measurements is below. Measurements originally in the metric system also have been converted to the English system.

1 *mu*	0.165 acre
100 *mu*	1 *qing*
6 *mu*	1 acre
15 *mu*	1 *ha* (hectare) or 2.47 acres
100 *qing*	1,650 acres
1 *li*	0.31 mile
1 *zhang*	10.936 feet
1 *jin* (catty)	1.1 pound
10 *liang*	1 *jin*

Dynasties and Other Historical Periods

Paleolithic	ca. 1.7 million–8000 BC
Neolithic	ca. 8000–2000 BC
Xia Dynasty	ca. 2200–1600 BC
Shang Dynasty	ca. 1600–1066 BC
Western Zhou Dynasty	1066–771 BC
Eastern Zhou Dynasty	770–221 BC
Spring-Autumn Period	770–476 BC
Warring States Period	475–221 BC
Qin Dynasty	221–207 BC
Western Han Dynasty	206 BC–AD 8
Eastern Han Dynasty	AD 25–220
Three Kingdoms Period	220–265
Western Jin Dynasty	265–316
Eastern Jin Dynasty	317–420
Southern and Northern Dynasties	420–581
Sui Dynasty	581–618
Tang Dynasty	618–907
Five Dynasties Period	907–960
Northern Song Dynasty	960–1127
Southern Song Dynasty	1127–1279
Yuan Dynasty	1279–1368
Ming Dynasty	1368–1644
Qing Dynasty	1644–1911
Republic of China	1912–1949
People's Republic of China	1949–present

Chronology of the Three Gorges Dam

1919 Sun Yat-sen proposes a dam in the Three Gorges to generate electricity to industrialize China and improve navigation.

1932 The Nationalist government, led by Chiang Kai-shek, begins preliminary work on plans for a dam in the Three Gorges.

1939 Japanese military forces occupy Yichang and survey the Three Gorges. A design, the Otani plan, is completed for the dam Japan intends to build after its anticipated victory over China.

1944 American engineer John Lucien Savage visits the Three Gorges at the invitation of the Nationalist government and completes a feasibility study for a large dam Xiling Gorge.

1946 The U.S. Bureau of Reclamation and the Resources Committee of the Republic of China sign an agreement to cooperate on the design and construction of the dam.

1947 Construction stops during Chinese Civil War.

1954 After devastating floods along the Yangtze, Mao Zedong asks the Soviet Union for assistance in planning and building the Three Gorges dam.

1958 Mao Zedong inspects the Three Gorges. Planning for aThree Gorges dam is stalled.

1969 Mao Zedong turns down the New Three Gorges dam proposal, citing national security reasons.

1970 The Chinese government decides to build the Gezhouba Dam in Yichang as a preliminary step toward a larger Three Gorges project. Work begins on the Gezhouba Dam. Construction falls into chaos.

1972 Construction of the Gezhouba Dam is suspended and does not resume until 1974.

1979 The Ministry of Water Conservancy submits another proposal to build a dam in Xiling Gorge.

1982 Deng Xiaoping announces his support for the Three Gorges Dam.

1984 The Chinese government decides to build a low dam in Xiling Gorge, creating a reservoir with a water level of 492 feet. Chongqing objects. The central government begins preparations to form a new province in the area, to be called the Three Gorges province.

1986 The State Council orders the Ministry of Water Conservancy to re-examine the project. The idea of establishing a new province is abandoned.

1989 The Yangtze Valley Planning Office, the government office in charge of the dam project, revises its earlier recommendation and proposes a high dam with a reservoir level of 574 feet. Construction of the Gezhouba Dam is completed.

1992 Despite ambivalence about the project, the National People's Congress passes the proposal to build the dam, with 1,767 votes in favor of it, 177 opposed, and 664 abstentions.

1993 The Three Gorges Project Construction Committee (TGPCC) is established to oversee the dam's construction. The Chinese government announces plans for a development-oriented resettlement policy.

1994 Premier Li Peng officially announces the start of the construction of the Three Gorges Dam.

1997 The China Construction Bank issues bonds to finance the dam's construction. Chongqing is elevated to the status of a municipality at the same level as Beijing, Shanghai, and Tianjin, reporting directly to the central government. The diversion channel, allowing work to begin on the main channel, is officially opened.

2003 The reservoir water level is scheduled to rise to 443 feet. Over 70 percent of the approximately 1.2 million people to be moved will be relocated prior to this time. The Three Gorges Dam will start generating electricity and begin paying for itself.

2006 The water level tentatively scheduled to rise to 518 feet.

2009 The Three Gorges Dam is scheduled for completion with a full reservoir level of 574 feet.

PREFACE

FOR HUNDREDS OF YEARS, THE PEOPLE along the shores of the Yangtze and in the mountains and valleys of the Three Gorges have owed their livelihood to the river and at the same time lived with the constant threat of having it taken away by the timeless cycle of floods and drought. The Three Gorges region, though isolated geographically from the rest of China by almost impassable mountains and the treacherous currents of the river, has also been affected throughout history by outside influences, from armies struggling for control as dynasties rose and fell, to foreign missionaries and the pervasive political and economic upheavals of the twentieth century. The local people have watched outsiders come and go, have had their lives disrupted or improved to one extent or another, and for the most part have gone back to life much as it always was, to farming or fishing or providing goods or services to the passing travelers or armies.

The Three Gorges Dam, now under construction in Xiling Gorge, the easternmost of the gorges, is another force of change brought to the area from the outside. Like many other historical events that the people here have witnessed, the momentum for the dam was not initiated locally, nor are its consequences under their control. The dam is something that they, as in the past, must endure, accommodate, make the best of, and profit from if possible. Unlike previous political or natural disasters, however, the dam will leave the physical environment of the entire region irrevocably changed, not only flooding cities and towns but also eliminating many of the cultural sites and markers in an area where the past is tied to the present by abundant historical sites and the tales of ancient heroes, goddesses, demons, and dragons that have been transmogrified into the very rocks and mountains that make up the Three Gorges.

In this book, I have attempted to provide a glimpse of the history and the current situation of a remote area, as the people who live here stand on the brink of immense personal and social disruption. Relying primarily on years of conversations with local people (many of whom have been referred to by names other than their own, at their request), county histories, and anthologies of local literature, as well as other Western Chinese historical and contemporary source materials, I hope to give a sense of how the people who have lived here see their past and future, inevitably filtered through my own experience of learning about it, one that now stretches back some twenty years. Written materials about the area often contain contradictory information, while the stories relayed by people have significant regional and personal variations and do not always bear great resemblance to what is believed to be historical fact. I have tried to present a picture that is accurate to the greatest extent possible, but the stories I tell may be less a representation specifically of what did happen than what people believe occurred and why that is important to them.

My own interest in the Three Gorges was sparked in the early 1980s when China was taking its first steps toward economic and political opening. I was a graduate student in anthropology at Columbia University, trying to decide on a thesis topic and where in China to do my field work. As a result of a chance meeting and the consequent, if not quite accurate, assessment that I was qualified to lecture on Chinese archeology, I was hired by Lindblad Travel, the great luxury tour company of the time, and soon became a frequent visitor to the Yangtze on their ships.

Eventually, I joined the U.S. Foreign Service and moved to Beijing in 1987, where I remained until the end of the summer of 1989. From there I was transferred to East Berlin, and three years later returned to the United States to cover Burma from Washington. While moving from one assignment to another, I remained drawn to the Yangtze River and the people on its shores, whose way of life seemed to transcend time. In the summer of 1994, I set off to visit some of the small towns in the gorges once again.

In my early trips to the Three Gorges, the dam had been nothing more to the ordinary people who lived there than a distant fantasy of some faraway officials in Beijing, but by the mid-1990s it had taken on a foggy reality. I returned to work at Harvard in 1996, and since then, during the years I have been at the Fairbank and Asia Centers, I have gone back to the Yangtze each year to learn more about the past and to watch the future unfold. This book was written over that period of time and reflects in part the final days of the old towns and communities in the gorges. As the book goes to press, many of these towns have already begun to disappear. Most of the residents of the old-fashioned town of Shibao Block are now living in the tall buildings covering what were fields of sorghum, much of Fengjie's city center has been dynamited, and the lower half of Wushan is almost gone. The old town of Zigui, in

Xiling Gorge, has disappeared entirely, its residents all relocated a few miles downstream. The first significant increase in the water level resulting from the Three Gorges Dam is scheduled for the early summer of 2003. By then almost all of the old towns in the Three Gorges will be demolished and cleared away, their inhabitants moved to higher ground.

In the course of writing this book, many people have generously provided advice and assistance, and numerous students have aided me in finding research materials and in plodding through the county histories of the Three Gorges. Out of these many people to whom I am grateful, I would particularly like to thank Melvin Ang, Holly Angell, Xiaoping Chen, the Chethams, Robert DiPasquale, Jun Jing, Madeleine Lynn, Tania Oster, Jonathan Schlefer, Judy Shapiro, Anne Thurston, Douglas Woodward, and Shuxi Yin. I am also deeply appreciative of the institutional and personal help from the Fairbank Center for East Asian Research and the Asia Center, and in particular to Merle Goldman, Nancy Hearst, William Kirby, and Ezra Vogel. I would also like to thank Florence Ladd and the Bunting Institute at Radcliffe College, the Millay Colony for the Arts, and the staff, management, and crew of Victoria Cruises, both in China and New York.

Many people before me have written books about the Yangtze River and the Three Gorges. For readers interested in learning in more detail about the river in the modern era I would highly recommend the following works in English. These have provided me with a wealth of information for which I am grateful. These include: *China's Warlords* by David Bonavia, *Yangtze! Yangtze!* by Dai Qing, *Szechwan and the Chinese Republic* by Robert Kapp, *Yangtze Patrol* by Tolley Kemp, *Policy Making in China: Leaders, Structures and Processes* by Kenneth Lieberthal and Michel Oksenberg, *Yangtze River: The Wildest, Wickedest River on Earth* by Madeleine Lynn, *Mao's War against Nature* by Judy Shapiro, *Yangtze: Nature, History, and the River* by Lyman Van Slyke, *The River at the Center of the World* by Simon Winchester, and *The Long Quest for Greatness: China's Decision to Launch the Three Gorges Project* by Liangwu Yin.

Finally, and most important of all, I am also deeply indebted to the many people in and from the Three Gorges region of the Yangtze River, who took the time to tell me about their lives and their thoughts about the past and future for without them this book would not have been possible.

INTRODUCTION

I FIRST SAILED THE YANGTZE IN 1983 ON THE *KUN LUN*, a foreign-leased cruise ship that went back and forth in seedy glamour between Shanghai and Chongqing. Over the next years, I made many trips along China's longest river, most often on its upper reaches, the section between western Hubei province and central Sichuan province in which the Three Gorges, a spectacular 120-mile stretch of mountains, ravines, and once deadly currents, are located. On board ship, I lectured to foreign tourists about Chinese history, shepherded them in and out of museums and factories, took them to hospitals, and dispatched them onward when necessary on stretchers and in urns. These long, leisurely, and contradictory trips, full of strange and fleeting intimacies, gave me my first introduction to river life.

Sailing along the upper Yangtze on a luxury cruise ship had its ironies. It provided a view not only of extraordinary natural beauty, but of day-to-day life in one of China's poorest regions, and of the mountain villages and crumbling ancient towns to which the ship's crew returned home every week. This was a world far away from our overstuffed armchairs and velvet drapes, one which few foreigners or city Chinese could begin to penetrate. Cocktails were served on the prow of the boat as the sun set over the river while old and new money from New York, Los Angeles, and points in-between sipped scotch and gin poured by girls from the hills of Sichuan. At dinner time, Lao Yu, a smiling man with no other known duties, went through the boat, calling passengers to the table with his small silver bell. Outside the dining room windows, fishermen and vegetable merchants in rowboats and sampans floated by, along with bits of rubble and tin cans, parts of houses, and drowned livestock. Anything not needed or wanted, from leftover food and garbage to large pieces of machinery and rattan couches, was tossed overboard or off the riverbank. "*Gei Changjiang chi!*" (Give it to the Yangtze to eat!), someone would say, heaving the object into the water from the shore or boat. Humans sometimes met the same fate. While serving dinner, the waitresses would cry "*si ren, si ren*" (corpse, corpse) every time a bloated, decaying body of a suicide, a villager too poor to be buried, or a newborn baby girl drifted by. The dinner staff would quickly analyze the cause of death and then avert their eyes, announcing it was bad luck to stare at the dead. Local officials reportedly felt it was equally bad luck to deal with these *shui da bang*, or water logs, as they were called, and rumor had it that corpses were only picked up if there was anything left of them by the time they reached Shanghai.

During daylight hours the noise on the ship never stopped. The young cleaning staff liked to work to orchestral versions of Mendelssohn's "Wedding March," and "Old Black Joe" blasted repeatedly from the central loudspeaker. A favorite for late afternoon mahjong sessions was the Boney M song, "Ra Ra Rasputin," popular in China in the early 1980s, if possibly nowhere else. Someone else would play the piano, usually a classical Western piece, while nearby the girls with the vacuum cleaner that sounded like an airplane hummed the latest Teresa Deng song from Taiwan or the Chinese version of "Red River Valley." Every so often the political commissar would call a meeting to discourage this unhealthy interest in decadent culture, and for a day or two only the loud and constant arguments about the price of green beans and fish would echo through the ship from the deck below, and then the Taiwanese songs would begin again.

At night the river was enveloped in darkness, broken only by the dim green fluorescent glimmer of distant villages and the sudden, blinding arc of searchlights circling eerily across the water. The cacophony of the day was replaced by the slurping sounds of the river, the sharp blast of ship horns, and the occasional singsong shouts of a harbormaster from his dock, yelling out the name of his dock and town through the heavy fog. On the shores behind the blackness of the night river were hundreds of villages and monstrous cities of concrete and foul-smelling smoke. They reemerged in the daylight, chaotic collections of stucco and sod houses in the hills, grim, gray apartment blocks next to factories spewing unrecognizable, boiling liquids into the river, throngs of peasants laden with baskets of green beans and eggplant, and laborers carrying coal on shoulder poles.

The Yangtze is one of the longest rivers in the world, flowing 3,900 miles eastward from its source at about 20,000 feet above sea level on the Tibetan-Qinghai plateau, where the Mekong and Salween also make their start, to Shanghai and the East China Sea. Like Gaul, it is divided into three parts, in this case, the upper, middle, and lower reaches of the river. The lower and middle reaches, from Shanghai to Jiujiang and Jiujiang to Yichang are generally broad and calm, the traditional rice basket of China. The upper reaches, from Yichang to the river's source, are dangerous and wild. Only the 1,700-mile portion of the river between Shanghai and the city of Yibin, 230 miles west of Chongqing, is navigable for commercial shipping. For over 2,000 years, the Yangtze has been the great transport route linking the coast with the west and southwest. It provided irrigation for the farmers who fed China, yet with terrifying regularity, it turned against them, sometimes drowning tens of thousands of people in days.

In the upper reaches, the river narrows and the countryside roughens into the spectacular and sometimes violent beauty of the Three Gorges, where the mountains rise into sheer cliffs and peaks above what were once treacherous shoals and rapids, now mostly blasted away. The Qutang, Wu, and Xiling Gorges have distinct geographic and historical characteristics but share a similar terrain and culture. Drunken poets, Taoist visionaries, and painters have retreated into this remote and beautiful landscape for over fifteen hundred years, as have invading and fleeing armies who struggled against the river and the impassable terrain from earliest times until the Second World War. The isolated and hard-to-reach towns within and beyond the gorges were places where officials were sent into exile to wait for the political tides to turn.

As far back as the sixth century, poets both elegized the beauty of the region and lamented being stuck here, usually focusing on the unkindness of fate and their loneliness and boredom. Over a thousand years ago, Bai Juyi, a scholar-official banished to what is now the county town of Zhongxian, west of the gorges, complained in his poems about the local conditions and people in the region, traditionally known as Ba, named after its original inhabitants. He wrote:

> Henceforward, I am relegated to deep seclusion
> In a bottomless gorge flanked by precipitous mountains . . .
> The inhabitants of Ba resemble wild apes
> Fierce and lusty, they fill the mountains and the prairies.
> Among such as these I cannot hope for friends
> And am pleased with anyone who is even remotely human. [1]

Much later, nineteenth- and twentieth-century Western merchants, missionaries, and novelists chronicled the harsh river journey inland and the extraordinary work of the trackers, naked men with harnesses and ropes attached to their bodies, who were responsible for guiding the steamships and large sailing ships through the gorges. This was a dangerous art and one of the few ways to earn a living for men with no land or education.

The steep banks of the Yangtze, though of enormous cultural and historical significance, have never been an easy place to live. The soil near the river is rich and good for cultivating wheat, sorghum, rice, and vegetables, but it is vulnerable to floods. The hillsides are famous throughout China for their tangerines, but these grow only at a narrow band of altitude, neither higher nor lower. In villages only a one- or two-hour walk away on the slippery narrow paths that wind through the high terraced fields, the mountains are higher and the soil poorer. Crops do less well, and it is almost impossible to transport anything more than you can carry on your back on the mountain paths.

Despite improvements in navigation and transportation, the area within and beyond the gorges remains remote in many ways. This is China's interior, home of historical and mythical kings, monsters like the *yeren* of Sichuan province (the "wild man," a half-man, half-ape creature somewhat similar to the abominable snowman), a region of conservative traditions and persistent independence from the rest of the country, where economic innovation and worsening poverty have existed side by side throughout the reforms of the past twenty years. Now, as a result of what will be the largest dam in the world, under construction since 1997 in Xiling Gorge, the easternmost of the three gorges, the area is also in the midst of social upheaval and massive change and may undergo more of a transformation within a few decades than it has in most of the past two thousand years. Though many travelers have depicted the hardships and frustrations of Yangtze voyages, there are few accounts of the experiences of the people who live in the towns or work on the river. The population displacement and flooding resulting from the Three Gorges Dam will largely diminish the opportunity to record the memories and history of this region.

Once the Three Gorges Dam is completed in 2009, the water level will rise as much as 350 feet in a hundred-mile stretch of the river, and create a 360-mile-long reservoir. The water will submerge over a dozen large cities, almost 1,500 villages and towns, and innumerable historical and cultural sites. Over a million people are being moved, voluntarily or otherwise, altering not only their lives, but the lives of a multitude of others whose existence is intertwined with the river. Age-old communities are being disrupted or dispersed, and a region already struggling with the impact of widespread rural migration is facing a greater need for labor to rebuild, yet has even fewer incentives to keep its youth in an area many wish to flee.

The communities on which this book primarily focuses, Shibao Block, Fengjie, and Wushan, are all located in Sichuan province on the upper Yangtze, within or beyond the Three Gorges. They typify the river towns of this area, rural communities undergoing sudden urban growth spurts, places that have remained largely isolated from and neglected by the outside world, but were nonetheless deeply affected by it. From the Sino-Japanese War to the Cultural Revolution and on to the present, local residents have witnessed political changes that disrupted or changed all of China and have had to cope with the outsiders who arrived as a consequence. At the same time, local customs and traditions in many cases have survived relatively intact. People still explain the origins of local mountains and rocks with the stories told by their grandparents, and girls marry out into the same villages as they did generations ago. Historical sites, which are key to the identity of almost every river town, continue to draw people in and provide a boost to the local economies.

The river towns have much in common, but their size, location, resources, and luck have also made their individual experiences unique. Shibao Block was a farming community of 2,000 people; Fengjie and Wushan were ancient walled cities. Traditionally, Fengjie was a prosperous commercial center; Wushan a port city of 70,000, contending with terrible poverty and sudden wealth. In the following pages I have attempted to tell the history of these and other communities, and the river that connects them, through the stories and historical records of the people who have lived there. Included here is also the story of how I sought out this information and of my interaction with the local people, for it reveals much about the interaction between this area and the outside world. The book concludes with a final look at these towns as they were before the Three Gorges Dam and a glimpse of their uncertain future.

CHAPTER 1

THE TOWNS

SHIBAO BLOCK HAS A KIND OF CHARM ONE DOESN'T SEE much in China. Narrow cobblestone roads are lined with two-story wood and stone buildings with graceful shuttered balconies. A hundred yards from the river, the main street, too narrow for cars, is a jumble of shops selling suitcases, shoes, chili, green peppers, fans, cold medicine, and clothes, scattered among hairdressers, restaurants, and barber-dentists still pulling teeth with filthy iron tools. Old people sit on stools outside their shops for much of the year, fanning themselves and calling to their neighbors across the ten-foot wide road. Their voices are loud across the short distance. Town residents comment upon everyone who goes by and everything that goes on. With a population of about two thousand, and only another thousand or so in the dozen villages within a few hours' walk, everyone who belongs here knows everyone else. Visitors are plentiful, but because they have no social or community connections, they make scarcely any impact except in the money they spend.

The Shibao Block Pagoda, a twelve-story red tower built on the side of a high cliff in the late eighteenth century, is what brings most travelers here. Looming over the town, the striking wooden structure houses several important Buddhist scriptures and was once known for its magical features. These include a hole in the stone floor at the top of the building from which rice flowed spontaneously. It is said that one day a greedy monk wanted more rice and tried to make the hole larger. The punishment for his gluttony was an end to the rice. The special powers of a second hole, located in front of the pagoda, are more questionable. This is the "duck-throwing hole" into which one could toss a duck and watch it reappear on the Yangtze, hundreds of feet below, only seconds later. The transgression that ended this source of amusement is lost to history.

Whatever moral lessons the tower may have had to offer, the fact that it was big, red, old, and visible from the river was enough to make its existence critical to the town's development and the focal point of contact with the outside. Shibao Block's experience is typical of many Yangtze communities whose historical sites guaranteed interaction with the outside world. Retreating armies, Buddhist scholars, Communist leaders, and ships full of tourists have stopped here. In the late 1960s, during the Cultural Revolution, when Chairman Mao declared all transportation free so that the youth of the country could bring revolution everywhere, the state-run East Is Red ship line, the only one on the Yangtze, brought students from all over China through the Three Gorges and the towns beyond. Red Guards from neighboring towns also made their way to the tower, some to see it, some to attempt its destruction. Placed under the protection of Zhou Enlai, the pagoda escaped serious harm, though the stone Buddhas once on its rooftop temple were smashed and stolen. In the 1970s the town had few visitors, and these were mainly students sent down to the countryside to live in villages as part of the Cultural Revolution policy requiring city youths and intellectuals to "learn from the peasants." Vacationing foreigners and southern Chinese began to arrive in the 1980s and their numbers have been multiplying ever since.

The town has two docks, one for high water, and one for when the river is low. When the water level is low, there is a long expanse of sandy beach that slopes upward to the steep concrete steps that lead to the town center. Peasants train young water buffalo to plow here, where there are no crops to damage. Down the beach, men repair rickety-looking boats and grandparents follow small children in circles. In the early morning the elderly exercise and the middle-aged fish, and at midday vendors of painted rocks and sedan chair bearers brave the brutal heat to await the arrival of the tourist ships. On summer evenings, when the temperature is often above a hundred degrees Fahrenheit, the entire town congregates by the river, the only place where the air moves.

The beach resonates with lilting Sichuan voices softened by the sound of the water and the thick humidity, which seems to swallow up other noises. People talk about how hot it is and whether someone has paid too much or too little for today's purchase, when a brother or a son is coming back from working on the docks in Chongqing or in the factories of Guangdong, and how much he is being paid and whether or not it is worth it. Like many conversations overheard

Figure 1.1 View of Shibao Block Pagoda from below

in China, these public discussions are often circular and almost a ritual—it's too hot, it's too cold, you're wearing too much, too little, you'll get sick, what are you busy with, where is your mother? Everyone on the beach, regardless of age, exchanges jokes and good-natured ridicule. They repeat movies and soap operas verbatim and report in detail to one another about any local altercation. Except for a few schoolteachers, usually assigned here against their will, and women who have married in from neighboring towns, everyone has known one another since childhood. They may not always get along, but they have always been here, and, for the most part, so have their parents and grandparents and everyone else before that.

The open space of the beach, lighted only by channel markers and passing ships but occupied by half the town, provides privacy for young couples and others in need of a place to discuss whatever they may not want the whole world to know. The only private place in China, it has been said to me repeatedly, is one filled with people, for if you are alone with someone, everyone knows you are up to no good and they will try to find out as much as they can. Only with dozens or, better still, hundreds of other people around are you not suspect. Here in public, men and women set up business transactions, plan marriages, and scheme to leave Shibao Block far behind. Though children chase after one another and old women knit and gossip, the night beach has its melancholy side. There are fewer twenty- or twenty-five-year-olds than one might expect, and those who are here wish they weren't. Young people stare wistfully at the passing ships and long for Zhongxian, the county seat that poet Bai Juyi loathed so much, with its two movie houses and karaoke bars. They dream of experiences and opportunities elsewhere, some past, some future, some totally imaginary.

The town shopkeepers and government officials and the farmers who live near the river are relatively prosperous for this region of the Yangtze. Life is comfortable enough, but it is a stratified society with all the intrigue and distrust that characterize small communities everywhere, and Chinese ones in particular. Complex obligations and opportunities for exploitation abound, and the level of suspicion and discontent that permeates life in the town and neighboring villages is high. The resulting frustration evident among the peasants and tradespeople contrasts sharply with the self-satisfaction characteristic of the town's few educated people and officials, whose minor power brings substantial rewards. One local irony is that while many ordinary workers spend their lives dreaming of ways to leave, the lucky few who do manage to get out and obtain an education often spend years struggling to get back to where they started. They now live a privileged life, a modern version of the scholar-official of centuries ago, divorced from the day-to-day problems of the rest of the population and unaware of the distance that separates them.

An ordinary day in Shibao Block begins before sunrise. By dawn, hawkers have positioned themselves on the stone steps leading to the town with baskets of fruit and lukewarm drinks they hope to sell to disembarking passengers before they reach the main street. Nearby, the elderly man who runs the candy and noodle shop in the ramshackle hovel where he lives with his daughter-in-law and two granddaughters is sweeping the dirt road. His son has gone to Chongqing to pay off the fine for the second child, who was supposed to be a boy and wasn't. In the town, people sit on their doorsteps and eat cold, watery rice for breakfast and brush their teeth in the side alleys. The wooden storefront shutters open and the business day begins. Women whose houses have running water hang laundry from their balconies, while on the outskirts of town, others carry their wash to the river. The fan maker takes out his basket of rattan and begins the day's weaving between puffs of smoke from an old-fashioned, foot-long pipe.

Shibao Block has several different rhythms to it, all traditional village patterns altered and adapted to a semi-modern life. The two major occurrences that mark time and produce activity are the market day, held once every ten days according to the lunar calendar, and the arrival of the boats, which varies with the season. The market brings the interior out, whereas the river brings the exterior in. On market days, peasants from villages scattered throughout the county set off as early as 2:00 AM for what can be a four-hour walk over the mountains. From the pre-dawn hours until about 7:30 AM, when the last of the latecomers arrive, the road leading into Shibao Block is filled with what seems like a massive exodus of refugees. Dressed in blue cotton clothes and black rubber boots, men and women pour into town with small children and vegetables strapped to their backs. Babies and chickens balance one another in baskets hung from the shoulder poles of adults and older children. For many villagers, vegetables are the only source of cash income. Money, not beans nor goods-in-kind, is necessary for cloth and tools and school tuition, but with the price of green beans about ten U.S. cents a pound and every farmer selling the same crop, most families earn next to nothing. The most motivated farm wives are in Shibao Block's town center by 6:00 AM, working hard to persuade the cooks from the government canteen and small restaurants that their vegetables are the freshest, their pigs, the tastiest.

The mood of the moving crowd is not exactly somber, but neither is there any noticeable sense of fun in taking part in the market day. People say this was different once, back when their grandparents were children and the alleys were filled with storytellers and performers. There are still a few itinerant fortunetellers and magicians around from time to time (a local numerologist, for example, told me not to travel at age fifty-two, and I have repeatedly run into a man breaking bricks on his head), but they are relatively rare. People are here to do business and in the first hour of the day there is a ferocious jostling for position on the streets and constant shouting about the quality of lettuce and tangerines. This subsides as the day goes on. By early afternoon, women are gossiping back and forth with relatives from other villages and tending to their own errands. By three or four, the villagers have packed up what is left of their goods to make the long trek over the hills before dark. The crowds on the road out of town are thinner, less energetic.

Like other rural areas of China, Shibao Block first went through land reform and then the formation of cooperatives and communes, which resulted in the closure of private shops and the eradication of private plots. During the Cultural Revolution, the villages around Shibao Block, which are hamlets of a few dozen people with the same surname, were turned into work teams and brigades. In the early 1980s, with the introduction of the "household responsibility system," families, rather than production teams, again resumed responsibility for most agricultural production. By the mid-1980s, the teams were back to being villages. Red Star Brigade, for example, was the Li Family Village once again. This change in structure, which provided rewards for hard work, as well as for good land and large families, allowed some to prosper while others floundered, and new groups of wealthy and poor farmers emerged.

The social and economic structure of greater Shibao Block, and the characteristic mix of hope, resentment, and resignation predominant among its inhabitants, are typical of many communities in the region. The people here are struggling to hold on to the life they know with the tenacity with which peasants cling to their land and to what is safe and familiar, yet they are also facing its imminent disappearance. This is not only because of the Three Gorges Dam, but the result of economic and political changes that have always caught Chinese peasants in their grips. Some of the river towns, like Xituozhen across the river or Wushan downstream, with its desperate energy, seem to suffer from a widespread exhaustion of spirit that is hard to define but immediately recognizable. It is a kind of weariness that tends to prevail in places where opportunities are few, nothing is clean, and there is a sense of decay and decline. Walking through some of the towns on the Yangtze reminds you of Walker Evans's photographs of rural, depression-era America in *Let Us Now Praise Famous Men*, a kind of dusty resignation that suggests endurance and possible survival, but little in the way of hope. From a single-child policy that is not consistent with family farming, to the fact that the government invested practically nothing in infrastructure development or industry in this region for decades, the people here are affected by a variety of issues beyond their individual control, to which they respond with widely differing capabilities and sophistication.

Peasants struggle both to stay and to leave for a variety of reasons, primarily lack of land or too much land but too few family members, or no other work, or simply profound boredom. Unemployment is widespread. In almost every family at least one member has gone south or west to work in factories or on construction sites. Some young people take their chances and go off on their own to find work in Chongqing, where the men end up as coolies unloading cargo in the harbor and the young women as child-care workers or waitresses if they are lucky, or in far worse situations if they are not. Most youths looking for work take the somewhat safer route of signing up with one of the factories in the Special Economic Zones in Guangdong province that use middlemen to recruit junior and senior middle school graduates every spring. Youngsters must prove they are in good health and unmarried, and then they are shipped off to light manufacturing factories, where they work for half the local wage but many times what they could earn at home. Their parents speak longingly of them, as if they were departed spirits who come back once a year for New Year's only to go away again. Some return home to marry and continue farming, or use the money they have saved to pay off taxes or invest in small businesses. Others join the hundreds of thousands of Chinese who have migrated illegally all over the country and will never be back. Older men go away to work, too, though they return after cold winters on construction crews in Xinjiang and Gansu and other far-flung provinces.

Just as market days demarcate the week, the landing of the tourist, transport, and cargo boats sets the daily schedule in Shibao Block. When the arrival of a tourist boat coincides with the market day there is pandemonium, with hundreds of Hong Kong Chinese, Germans, and Japanese pushing their way through the eggplant to the pagoda and getting lost in the side streets. Though the tourists provide almost everybody here with an opportunity to supplement their income, the townspeople hate it when the peasants and the tourists swarm through together, everyone in someone's way.

Shibao Block now gets about 100,000 visitors a year, of whom about 40,000 are foreigners and visitors from Hong Kong who arrive on the tourist ships. The other 60,000 are domestic travelers who come by bus or boat, often as part of government-organized trips for students, state workers, and old people. Despite this huge influx, they are in town at most a few hours and have little significant interaction with local residents. The impact of tourism on the overall economy has been limited, for despite the number of visitors, all return to their boats and buses after a two-hour stop, and as a result, there has been no development of hotels or a travel-related industry. The sightseers have nonetheless provided a market for the entrepreneurial efforts of individual households and shopkeepers. On days when the big ships dock, the stores close their doors for a few hours and the owners cram the streets with folding tables stacked with painted rocks, souvenir key chains, postcards, sunglasses, jade balls, rolling pins, umbrellas, plastic towers, hundred-year-old coins, shoes for bound feet, and other relics of the past and present. Some of the wares are homemade or have been found in old trunks and cabinets. Other goods are imported by energetic salespeople from Wanxian, the closest big city, an eight-hour boat ride away.

The sedan-chair business has also had a revival. Under Communist rule, sedan chairs and rickshaws were viewed as examples of feudal exploitation and banned. Except for the occasional rural wedding or for trips to the hospital, they were rarely seen for nearly forty years. Now almost every family in Shibao Block has one. Town residents build them in their spare time for the tourist trade, attaching anything from kitchen chairs to fanciful yellow thrones to bamboo poles strapped together with twine. One of the many current social contradictions is that services and traditions eliminated by the Communists are back as status symbols and luxuries, heartily approved of by both the people who want them and the individuals providing them. It is the foreigners who complain that it is wrong for hard-working peasants to carry heavy Americans and Cantonese up the steep slope in hundred-degree temperatures, and the local Chinese who say foreigners are crazy and the ones suffering from "old thinking." This is legitimate work that brings in good money, not exploitation.

Officials say that the revival of the sedan-chair business demonstrates initiative and self-reliance and is an example of traditional culture, but it has also led to new exhibitions of aggression. To curb the violent arguments over who can station his sedan chair closest to the dock, the town has instituted a licensing system and threatened to impose a fine for brawling, but this has had little impact. The sedan-chair bearers still rush toward disembarking passengers, pushing them or dragging them toward a chair by their clothing and yelling angrily at those who want to walk. The best targets are young and middle-aged men, mainly southern Chinese, Koreans, Japanese, and Germans, who tend to want their photographs taken as they are borne up the hill by girls and old men. At the end of the road near the pagoda, whether by design or miscommunication, more trouble erupts when a passenger discovers that the price of the ride has doubled. Sedan chairs are flung to the ground and high-pitched, hysterical screams resound through the quaint streets.

Just as on market day, the streets are empty again by mid-afternoon. Everyone who doesn't belong in the town is gone, and everyone who does is back at work. In addition to agriculture and the various economic opportunities generated by the pagoda, by the mid-1990s, Shibao Block had thirty-seven storefront medicine shops selling a combination of Western antibiotics and Chinese herbs and traditional remedies, six clinics, a hospital, an old-age home, and several blocks of cloth merchants and tailors. An elderly undertaker still embroiders lotuses (symbols of rebirth and purification) and bats (symbolizing good luck) on footrests for the dead, and his staff makes masks of dragons and folk heroes such as the Monkey King for opera performances and New Year's celebrations. The town has a barrel maker, a brick kiln, a few stationery shops, many government offices, and two factories, one of which produces Zhongxian Baijiu, a locally favored brand of sorghum-based grain alcohol, the other firecrackers. Plans have long been underway for a government-owned plastic tarp factory and other small new enterprises. The idea is that when the entire town is moved to its new site, a few miles up and back, projects of this sort will help in the transition to an economy in which light industry plays a greater role.

Though the general public of Shibao Block is far more concerned with the present than with the future, an awareness of imminent change is always there. In casual conversation, people on the streets point to the markers that indicate where the water will be in a few years, and say, looking around them, "This will all be gone," and often adding without any obvious distress at the thought, "and so will I." For years, the residents of the Three Gorges region viewed the possibility of a new and massive dam much like the threat of an earthquake or a volcano—a potential disaster, but one without any reality, like any other looming catastrophe, whether natural or government inflicted. Most local people perceive the dam and the rise of the river as something about which little can be done, except to wait and deal with whatever happens when the time comes. At the same time, no one is naïve about who stands to lose and who will gain, or where each individual stands in that equation.

Once dusk falls, there is little to do in Shibao Block. The dozen or so restaurants, which serve mostly the owners' family and friends anyway, close by seven. The movie house, with its steady run of Hong Kong martial arts films, is open only on weekends. A disco, which served the younger set of the town, was located on the second floor of an old warehouse and used to be open

Figure 1.2 Textile and basket vendors outside Old People's Home in Shibao Block

until ten, but that has been closed, supposedly because it attracted a ruffian crowd. The Recreation Center for Retired Cadres, however, is still open until eight and is filled with elderly party members and officials playing mahjong and cards beneath pictures of Mao, Lenin, and Stalin. Though the town has been electrified for a quarter of a century, most homes and shops are lighted by a single bulb or fluorescent tube, and in the stifling heat of the summer, there is little incentive to be at home awake. A few of the private shops that sell snacks and sundries have television sets that draw in passers-by to watch a fuzzy version of the latest soap opera broadcast from Chongqing. Unlike many of the other river towns, or even some of the surrounding villages, there are no satellite dishes in downtown Shibao Block, and local familiarity with the outside rarely goes beyond the next few county seats.

⟡ ⟡ ⟡

There is a saying in Chinese that means "hard places breed hard people." Wushan is one of those places. With a population of 70,000, it is the largest city within the Three Gorges, one with a rough past and an uncertain future. Unlike Shibao Block, with its quaint cobblestone streets and traditional architecture, or once thriving Fengjie, Wushan combines the harshness of rural Sichuan poverty and its legacy of isolation, unemployment, and malnutrition with the squalor of a medieval port city. Located at the mouth of Wu Gorge, the second of the Three Gorges, the town is far from anywhere, yet it has been important in history from time to time because of its position at the confluence of the Yangtze and the smaller Daning River. Known for its own set of three gorges, the Daning flows northward toward the interior of China and provides a route inland to a mountainous region largely accessible only by footpaths.

Wushan is ranked among the poorest counties in China by both the Chinese Ministry of Agriculture and the World Bank, a list that includes places with a large proportion of people whose income is less than one U.S. dollar a day. Along the river, being poor is not much of a distinction, but Wushan is known for other things, mainly for what the Chinese call being *luan*, or chaotic and disorderly. This can mean many things, but here it includes the open prostitution and close-to-the-surface anger that is manifested in frequent street fights and tearoom brawls. Wushan is the only town in the gorges that can boast of several murders in recent years, and it is one of the rare ports where captains sometimes restrict their crews to the ship while docked.

Wushan's reputation as a harsh place extends to other spheres as well, including politics and economics. Many people have prospered here over the past fifteen years, but the divide between those who have done well and those who have not is deep and clear. In the mid-1980s, when passenger ships first began stopping on the edge of town, people from miles inland flocked to the riverbank to sell eggs, the only easily portable item of interest. Local peasants, with their heads wrapped in turbans made of dish towels instead of the traditional cotton strips, stood listlessly on the shore with their baskets. Lethargic children with scabs, shaved heads, and runny eyes hovered nearby. At that time the price of eggs was ten fen a catty (about four U.S. cents for around twenty eggs), a tenth of what they cost in Chongqing, but no one shouted out that prices were cheap or tried to undercut their neighbors. The peasants squatted silently on the shore, smoking and staring at the river, waiting for customers to approach them. Conversations were hard to start, as if the gulf between the villagers and anyone who came from the outside, even a few miles up the river, took just too much energy to bridge.

Wushan today bears little resemblance to the desolate spot it was a decade or two ago. These days it has the feel of a border town, a seedy place filled with men looking for something to do. The first thing you notice on the main street leading up from the harbor are the town's sixty-four hairdressing salons. Young girls and women sit on three-legged stools under twinkling Christmas lights in open-front shops and call out, usually without much enthusiasm or interest, "Gentlemen, get your hair washed, have a massage, gentlemen, come in, come in. Wash your hair, ten yuan, ten yuan." Behind them, on the wall facing out to the street, are eight-by-eight foot blow-ups of naked Caucasian women. Each shop has a different poster. There are pictures of reclining, standing, and coiling women, as well as a few having their clothes torn off by equally unclothed men, all of whom have massive amounts of coifed hair. The photos, which would not be allowed in public in much of the Western world (nor anywhere in China a few years back, when statues on bridges and Tibetan gods still had to wear skirts), seem to shock only the visitors. When I asked some local people what all this meant, were these all brothels or wasn't it a little strange to have the main street dominated by enormous naked women, they replied with a blank stare. "The pictures are very fresh and modern," someone would finally say, occasionally adding as a well-meant compliment, "They look like you."

When darkness falls, vendors set up tables and sell dumplings and *huoguo*, a local chili-laden dish of vegetables, meat, fish, and fowl boiled in broth, like fondue. With the lighted carts and hawkers' calls, the town has an almost festive air. There are new items for sale on the street—dresses from Chongqing, sweaters from Wuhan, shoes from Shanghai, and, once, gerbils from San

Francisco, via Wanxian. Imported by a young man with an eye for novelty, the strange-looking animals (advertised as pets and not dinner) drew a crowd that stopped traffic for a block. The mix of money and strangers on the street does not always go smoothly, though, and the carnival atmosphere sometimes gets surly, mostly a result of disputes between local people and transients over the prices of food, drink, hotel rooms, women, and anything else one might pay for. A government complaint bureau that was set up in 1997 received over 800 letters and phone calls within the first three months, reporting hundreds of ways of being cheated.

Wushan's main streets are named after the twelve fairies who helped the mythical Emperor Yu stop a great flood somewhere around the time of Noah's ark. After helping sweep the waters into a deep crevice that the legendary emperor had made for this purpose, the exhausted fairies turned to stone and became the twelve peaks of Wu Gorge. Since that early battle with nature, the elements have never let up. The county archives record six kinds of disaster—droughts, floods, fires, famines, avalanches, and ice storms—beginning in 200 BC and continuing until the present with dismal regularity. Fire destroyed the city three times, the townspeople starved to death or were eaten by wolves eight times, and the town has had scores of floods, great floods, and catastrophic floods.

The struggle to cope with such events and to earn a living still dominates the lives of the people here. Men and boys congregate every morning by the steps to the pool hall, where the road starts to slope down toward the river, to wait for day jobs hauling cargo. Passenger and freight ships make over 4,000 stops in Wushan each year, so there is a steady demand for people to carry things, but there is also no shortage of young men without regular work. Laborers may stand around for hours before they are hired and will earn little when they are. On the riverbank below, men who have come in from the countryside load coal shipments from the interior onto the transport barges that go on to Chongqing and Wuhan. Like the group outside the pool hall, they have few choices, but they are older, from poorer villages, and more in need of work.

By 5:00 AM, hundreds of coolies form a chain circling from the heaps of coal toward the waiting boats. (The word *ku* means bitter or hard, *li* is strength, and together *kuli*, or coolie, means bitter strength, and is the term widely used in China for laborers.) Each man carries two wicker baskets filled with soft coal, suspended from his shoulder pole, a weight of 150 pounds or more. After the men empty their loads onto the barge, they run back to the mountain of coal on shore, put down their empty baskets, and hoist up two more already filled with coal by other workers whose task it is to shovel all day. They then lift the baskets back up and sprint toward the river again while keeping pace with the men ahead of them. They will do the same thing over and over again until the morning is gone. A day's hard labor can bring in twenty yuan needed for building materials, clothing, or other things that do not grow on a farm. Women and girls also carry coal along the Yangtze, but not here. Great speed and stamina are required, and the women neither desire the work nor are they welcomed. Instead, they stay in the fields, which they care for alone when it is not planting or harvest time.

In summer hot pink is the favored color for men's undershirts. From a distance the brown riverbank is dotted with tiny brilliant specks. Looking at the shore from the harbor, behind these dots one sees a maze of red, white, and blue striped shacks winding across the rocky beach and onto the steep and wide staircases that lead to the road above. A climb of hundreds of steps is the only way to reach most of the towns along the Yangtze. The normal difference between high and low water can be as much as 250 feet in the course of a year, and more when it floods, so the towns are built high above the river. The stairs provide the only route up other than clambering over rocks and bramble or deep mud and slime, and like the streets of the town the stairs are lined with food stalls, repairmen, and souvenir stands for arriving passengers. Many peasants have left the land over the past few years, and farm families now struggling with urban life have joined impoverished out-of-town laborers in living on the steps to the river. Along the riverbank, the building material of choice and necessity has become the striped plastic sheeting used in Hong Kong for the cheap and indestructible bags into which emigrants and shoppers can fit a lifetime's belongings. Sheets of plastic are draped over a basic bamboo frame and secured with river rocks. A house, or at least a lean-to, can be put up in less than a day.

The people who live by the docks are the poor but not yet destitute, the enterprising of the down-and-out of Wushan who have set up tiny businesses of their own. The six families who have settled on the steps to dock number three, slightly away from the center of the harbor, are typical of many others who cater to travelers and boat crews. The man with the best spot on the steps, at the top by the road, is a middle-aged shopkeeper whose wife and children still farm in the countryside. He is doing well for himself with his concrete floor, VCR, and wooden lounge chairs, but aside from him, everyone else who lives and works on this set of steps seems to have ended up here because of some major or minor tragedy. On the top left of the stone steps is a couple whose farmhouse burned down. They came to the dock in hopes of earning a cash income so they could rebuild their home. The man across the

Figure 1.3 Wushan Harbor

way lost his temporary factory job and does not want to go back to his home village. Another family faces exorbitant fines, one for a truck accident and another for a second child, burdens both equally resented.

The couple saving for a new house runs a small restaurant. The wife cooks on a coal burner and her husband sells soft drinks. The family has a small table, a bench, and a stool on the packed earth floor, some cots in the back. Most of the day they sit on the steps, fanning themselves and snapping beans, talking to the neighbors, and moving back inside their shelter when it gets too hot or rains too hard. The Sichuan climate is a harsh one, good for growing crops because of the heavy rain and strong sun in summer and relatively warm, damp winters, but it is not a pleasant one for humans without heat or windows. In a tarp house, the water leaks in when it rains and the dirt floor turns to mud. Despite the lack of amenities, their son graduated from senior high school and is going to the prefectural capital of Wanxian for a year-long accounting course. His ten-year-old sister also plans to go to Wanxian someday. She studies hard, already knowing that this is probably the only good way out.

The boatbuilder and his wife live a few yards below. The wife serves me hot sugar water boiled with iodine to guarantee it is safe to drink. As we sit underneath the wooden skeleton of a small fishing boat, she offers me her seven-year-old daughter to take to America to educate. I am not sure if she is serious, but the child looks panic-stricken. Her mother asks how much it costs a foreigner to get a baby girl from an orphanage, a question brought up by almost every woman in Wushan, all of whom are aware of the number of baby girls now in Chinese orphanages as a result of the one child per family rule. When I ask them what they think about foreigners adopting these Chinese children, they tell me, somewhat to my surprise, that it is a good thing. One woman mentions a baby left at a crossroads just outside the town a few months earlier, someone else talks about another found in a basket on market day. Almost half the town is paying for a second daughter, and no one discusses whether there was a third.

Far above the riverbank, also visible from the river, is the new town of Wushan, where the city residents will move when the Yangtze waters rise. Located on the highest hill in the city, it shimmers in the distance, a brilliant white in a town where almost everything else is gray and brown. In Wushan, when people talk about the future, they point upward, gesturing vaguely in the direction of what will be left when everything else is underwater. Funded by different agencies with competing goals at the local, provincial, and national levels, the building of the new town has brought out tensions and conflicts that have existed within the community for decades. The new homes and offices, better looking than anything else that currently exists in the Three Gorges, are not for everyone. The apartments here are being assigned the old way—by work-unit—which means that each institution decides where its employees will live within the housing assigned to it by the government. This gives rise not only to fears of favoritism on an individual level, but to unending negotiations and scheming about which work-unit receives which blocks of housing.

The headquarters of the Communist Party, still an important force in the city, has an impressive compound on the main street and is moving to an equally central location. The Catholic church, on the other hand, is repeating the battles it has fought for the last half-century. The massive cathedral that once stood overlooking the river was torn down in 1954 at the height of the land redistribution movement and was replaced by an electric light bulb factory. After 1983, when property rights were restored to the Catholic Church in Wushan, the factory was closed and the slightly refurbished building served as a church for more than fifteen years. Locked in a series of disputes with the local government, no one was sure for some time where the church would find a new home when the town moved to its new site. Schoolteachers, already poorly paid, fret about where they will end up, and *jumin*, or ordinary citizens, who have taken the risk of leaving a job that provided housing or who never had found work in the first place, shrug their shoulders and say they'll see.

For what might be called the middle class of Wushan, the educated and salaried people who work for the government, man the hospitals, write the newspapers, and run the hotels, the new city on the hill represents a future they want. It means that change in China is not just something that they see in a snowy TV newscast from Beijing but, they hope, an apartment with indoor plumbing and jobs for their children that might stop the exodus to Wanxian and factories in the south. It means, the wheeler-dealers think, a time when the houses they build will be sold on a private market and there will be people who can afford to buy them, and there will be an express bus to the center of Chongqing.

⟡ ⟡ ⟡

Fengjie, one of the few places within the gorges with its city wall still largely intact, is beset by the problems of economic restructuring and the closing of state-run factories facing all of China, but it is doing well enough. Located by the mouth of Qutang Gorge, it is a place with a vivid history, which has gone back and forth from prosperity to poverty to the ordinariness of small-town affluence. Rich in salt and coal, commodities important for both political and economic reasons, Fengjie has been an important commercial center since as early as the fourth century BC, when its growth was spurred mainly by the salt trade. Sichuan was once an inland sea and when the waters receded during the Triassic era 250 million years ago, underground brine wells and solid salt deposits formed along with large pockets of natural gas and coal from millions of years of rotting vegetation. In the nineteenth century, the hundreds of brine wells in Fengjie each produced an average of 132 pounds of salt per day.[1] During the winter months, when the level of the river was low, men would dig deep wells along the banks of the river and then boil the brackish water over coal stoves. This cottage industry was one of the main sources of revenue for the region, bringing in, by some estimates, as much as two million pounds sterling per year, the equivalent of U.S. $124 million in 2001.[2] Coal and industrial products are still the main stays of life in Fengjie. Huge clouds of white smoke burst forth from gray factories while sickly yellow foam comes from others. The riverbank is coated with white lime, and clusters of men shoveling white hills of powder alternate with men shoveling black coal.

As in other river towns, long stone steps lead up to Fengjie, but here they are in better condition, wider and more elegant, and end at the main gates of the city. Farther from the center, smaller gates open up on winding steps and back alleys. In the 1960s and 1970s, when private shops were prohibited, the narrow lane just behind the city wall stood empty. The alleyway is now as packed with shops and peddlers as it was in Fengjie's heyday a hundred years ago. In the late nineteenth century, Fengjie was a raucous transshipment port where merchants had to pay taxes on their goods before proceeding farther in either direction. The town was also a resting place for travelers who had made it through the gorges. Trackers waited on the streets of the town for work pulling cargo and passenger boats back through the rapids, and captains restocked and battled over prices and taxes.

Like Wushan, Fengjie is a county seat, but a more orderly one, a city that residents like living in. People speak glowingly, if perhaps not altogether accurately, of the cooperation between the Communist Party and the companies working with foreign firms to build small power plants and new chemical factories. Though there is talk of corruption here, as everywhere in the gorges, there seems to be less antagonism between the city residents and the authorities than in many river towns. The wall circling the city has been transformed into a massive advertising billboard with twenty-foot posters for Panasonic washing machines and Sanyo rice cookers. Small shops and restaurants are flourishing, their windows filled with Western-style birthday cakes and wedding gowns, symbols of luxuries that are now commonplace in the city. Japanese gadgets and

Figure 1.4 Steps to the Yidou Gate of the city wall in Fengjie

appliances sell quickly in the department stores, while outside on the streets there is no lack of supply or demand for more traditional items. Mountains of ground chili are available on every corner in a spectrum ranging from brilliant red and orange to deep green, each powder of a slightly different hue, smell, texture, and taste. In the large public square, children and parents circle in hired electric bumper cars shaped like panda bears, lions, and other animals. Boys play kickball, women repair handbags, and the blind offer massages.

Near the city gate, large banners with pictures of dead rats lead the way to the exterminator's stall in an open-air market overlooking the river. The proprietor explains his products carefully, pointing out a poison much better than any you could buy in Chongqing, certainly better than anything available in the United States, because it kills rats very quickly, very dead. He has a wide range of traps, some cleverly designed in the shapes of mice or cats. Throughout the market, new thermos bottles, high-heeled shoes, clothing, pomelos, pears, and other items are spread out on the ground. Prices are high here, clothing somewhat more fashionable than farther into town. This is where travelers shop while their boats are in port, at least those who have the stamina to climb the 300 steps to get here.

Most of the city wall now standing was reconstructed a few years after the original fifteenth-century wall was destroyed in the great flood of 1870, the largest flood in the past thousand years. On the main, or southern, gate are inscribed the words "Looking Far and Wide," a quote from a poem by Du Fu (AD 712–770), the Tang dynasty poet who lived in Fengjie from 766 to 768. Du Fu had fled to Sichuan from Gansu, where he had been serving as an official, to escape the chaos caused by the Anshi rebellion in northern China in 759. He wandered from place to place for ten years, staying in Yuzhou (now Chongqing), Zhongxian, and Yunyang before ending up in Kuizhou, which is now Fengjie. Though miserable in exile and distressed at the state of political affairs that brought him to Sichuan, he wrote nearly a quarter of his life's work, or about 400 poems, during his two years in Fengjie. Many of his poems drew on local history, particularly the battles of the Three Kingdoms period (AD 220–265) when the military commanders of the kingdom of Shu were based in Fengjie. Near the city gate is a terrace lined with wooden lawn chairs and tables overlooking the mouth of Qutang Gorge and Kui Men, the massive walls of rock shaped like gate doors at the entrance of the gorge. The wooden chairs are rented out by the hour and offer a serenity rarely available in a place where travelers customarily crowd together in the heat and rain on the dock until they stampede to the gangplank. Though the fee is low, only the seemingly well-to-do are willing to pay it. A few young businessmen, in black leather jackets despite the heat, pour out glasses of grain alcohol and toast one another, snapping their fingers for emphasis. Three ancient brothers, the eldest of whom ended up in Taiwan with the Nationalist Party (KMT), repeat the current hour and the expected arrival and departure times of their ship like a mantra, while nervously tying and retying their bundles with pink plastic string. The men tell me which of them is the first-born, who second and third, and then ask me to guess their ages. I try seventy-five for the oldest, but he is seventy-nine and looks like a hundred. They tell me that when Lao Da (literally "old big," as all oldest siblings are called) came back to visit from Taiwan for the first time a few years ago, he brought them a television set, but it did not work in their village, a three-hour bus ride from Fengjie. After not seeing them for forty years, all he could bring was a broken TV! Now they have a satellite dish and the television works after all. They chortle at this and tell me once again to guess the ages of brothers number two and three.

Inside the city wall, the center of the town winds upward, the streets crowded with neighborhood markets and restaurants. Fortunetellers, mainly palmists and numerologists, have made a comeback and cluster by the main gate. The hospital is another favored location for fortunetellers and masseuses. Using the date and hour of my birth, a man with an abacus, his head wrapped in the traditional cotton turban once common in this area, calculates my good and bad years to come. The chart spikes up and down like a cardiogram, finally indicating that I should last until about eighty-five, a good age to die, which is what he tells everyone. I ask him if predicting the future is a family trade, but he snorts at the idea. Hardly, he says. He was a peasant and had worked in a small factory where he studied fortunetelling in a book in his spare time. It is more interesting than being cooped up inside and not so hard as field work. What's more, you can set your own hours and help people with their lives. The numerologist, addressed by people on the street as *shifu*, or master, the polite terms for drivers, cooks, tradesmen, and some teachers, explains that every day he has at least a dozen or so clients, and two or three times that many when the tourist boats come in. Most of the sightseers who stop by have their fortunes told for fun, but here in town he performs a real service, advising his regular customers or people worried about their children or money about what to do.

In addition to traditional trades such as boatbuilding, for which Fengjie has long been known, the town has become a regional industrial center producing fertilizer, cement, and silk. Not far from here, smaller dams have been constructed on the tributaries of the Yangtze, projects in isolated spots in the interior of Sichuan that evoke no controversy and are paid little attention

Figure 1.5 Loading coal in Fengjie

within China and almost none internationally. Fengjie was one of the first cities along the river to complete large sections of its new town. Here, like everywhere, there has been controversy over who will move where, but within the city at least, most residents and institutes initially seem relatively satisfied with the process, a fact that city officials attribute to having enough space and beginning planning early. This orderly progress was thrown into chaos in late 1998 with the discovery that an almost finished part of the new town close by the existing city was built on sandstone too soft to support the construction and was dangerously unstable. It had initially seemed that it would be necessary to take down all the completed housing and office buildings, which had cost an estimated 50 million yuan (approximately U.S. $6 million). Instead, the number of planned high-rise buildings was reduced from one hundred to thirty. Though concerns remain about the long-term stability of the mountainside, town officials seem confident about the safety of the construction and financial wisdom of their decision. Meanwhile, an additional site for the new town was chosen almost six miles away, and the population of Fengjie is now being redistributed to three separate locations instead of one.

While they wait to go, the townspeople are making the most of what will not be there later. Baidicheng (White King City), an important historical site from the Three Kingdoms period, is adjacent to Fengjie. Accessible until 1994 only by boat, the mountainside is now connected to Fengjie by a long bridge, built just before the beginning of construction of the Three Gorges Dam to make transportation for townspeople and tourists easier. One of the local ironies is that the graceful bridge, which stretches into the open air hundreds of feet above the river, will be under about seventy-five feet of water when the reservoir rises to its full depth in 2009. This, like all other structures, will have to be demolished to clear the way for navigation. The top of Baidicheng, with its temple and commemorative halls, will be an island. A short trip across the Yangtze is a cave, once used as a shelter from Japanese bombs, that the Fengjie branch of the China International Travel Service turned into a local version of a campground with pup tents outside, pool tables and a snack bar on the ledge, and a karaoke hall deep in the darkness of the interior. The hope had been to make a profit before the cave was submerged by the new dam, but the site was badly damaged by floods in 1998. On the opposite shore, an impressive pagoda-shaped tea house still stands. It was built originally as a viewing station for high-ranking observers when Jay Cochrane, a Canadian tightrope walker, crossed the Yangtze on a 2,000-foot steel wire strung 1,250 feet in the air. Invited by the Ministry of Culture, he set a record for the highest and longest tightrope walk, an attraction that drew over 20,000 visitors, including Zhu Rongji, later appointed premier of China.

The next such event was supposed to be Chinese twins bicycling across the river on the tightrope, but according to one young woman working as a local guide, the arrangements proved more difficult than expected, and the spectacle was postponed indefinitely. "Fengjie is still pretty boring," she added, "but there are lots of ways to make money." This, more than anything, is the driving force of the young people who stay in the city. The countryside is poor and state-run factories have laid off employees, but unemployment is still relatively low in comparison with most other small river cities. Government work-units and a few entrepreneurs continue to build hotels and restaurants below the new high-water mark, calculating that they will make back their investment before the final flooding in 2009.

The vast amounts of money that have come into the area to rebuild the towns and relocate the population have brought opportunities, legitimate and otherwise, for fast-thinking, clever people. After the quick money is gone, the question for Fengjie and other towns and villages along the upper Yangtze will be whether the abrupt transition from past to future can bring development and a better life without completely destroying the physical environment and cultural heritage. Fuling, the capital of the ancient Ba Kingdom, Fengdu with its ghosts, hideous and crumbling concrete Badong, and all the other towns and cities between the dam site and Chongqing share a similar past and face comparable losses. The Three Gorges Dam has brought in new roads and airports, high-rise buildings and factories, but it will result in new ugly places as well, and the river will be a long-ago memory no more real than the city wall of Beijing and other lost monuments. The government travel service along the river, which necessarily takes a positive view of what is to come, says the flooding will not diminish the beauty of the Three Gorges, it will make them bigger and better. Instead of a narrow river, there will be a great lake. Temples once too high in the mountains to reach will be accessible and tributaries too shallow for tourist boats to navigate will be broad and deep enough for cruise ships. Thousands of people will be able to lunch in new restaurants in the interior, and the guides will explain what used to be here.

CHAPTER 2

HISTORY AND MYTH

EVERY ROCK AND MOUNTAIN ALONG THE UPPER YANGTZE is here for a reason that can be explained. A fairy turned into a rock or someone's head was cut off and the dripping blood bored holes into the mountainside. The tales are part of the day-to-day life of the river, never far away from the present, and are so deeply ingrained that people of any age can tell you the most complicated stories of romances and battles of a millennium ago as if they had been there themselves, much the way some TV fans can recite the latest development from a familiar soap opera. Legends from the upper Yangtze tell first and foremost of the struggles between people and nature, with a combination of human and supernatural forces usually coming to the rescue. The tales are mostly explanatory, describing how a place came to exist or was named, but others glorify the heroes of a particular era or the virtues of a noble man. Few tales are cautionary or offer a moral, for though the wise and honest are portrayed as deserving of respect, they do not necessarily end up any better off than the duplicitous or crafty. Sometimes it is clear that behaving badly does not pay off in the end, but often it makes no difference at all.

As in many cultures, the history of the Yangtze River begins with a great flood and then salvation for some. Water is of key importance, both in myth and in reality. The Qin, the first dynasty to unify China, succeeded in doing so in 221 BC in part because of the extensive network of canals and irrigation that it had established in Sichuan. In China, with its far-flung population and large areas of inhospitable terrain, the harnessing and distribution of water remain key elements in the political and emotional history of the nation. This has always been of utmost importance along the Yangtze, where floods and droughts are a constant part of every life.

While it is an integral part of local mythology that Emperor Yu created the Three Gorges in ancient times by forging a great ditch to divert the flood waters, geologists take a different view of how they were formed and claim that they resulted from the shifting of the earth's crust during the great continental movements over two million years ago. Likewise, while local folk tales explain how the area came to be populated and communities formed, archeologists and paleontologists focus on different theories about the origins of life and civilization here. Some of the earliest human remains in Asia have been found near Wushan. Peasants often gather fossils, known as dragons' teeth and bones, from caves as ingredients for traditional Chinese medicine. In 1978, a farmer reported to local authorities the existence of fossils in Longgupo Cave (Dragon Bones Cave) near a village a short distance inland from the south bank of Wushan county. In 1984, scientists from the Institute of Vertebrate Paleontology and Paleoanthropology in Beijing and the Museum of Natural History in Chongqing surveyed the cave.

Over the next four years, researchers found over 10,000 teeth and bones from more than a hundred species of mammals. These included Gigantopithecus, a giant herbivorous ape related to the orangutan; the extinct mastodon elephant Sino-mastodon, an early species of horse; and Ailuropoda microta, a micropanda which is the ancestral form of the giant panda that is still indigenous to eastern Sichuan. Scientists also found a tooth and part of the jaw of a hominid that they determined to be between 1.79 and 1.96 million years old. In 1995, a team from the Museum of Natural History in New York visited the site and corroborated the age of the remains at about 1.9 million years, roughly the same age as the Java man specimens from Modjokerto and Sangiran in Indonesia. Chinese researchers originally assumed that Wushan man, as they called this example of early man, was a variant of Homo erectus, like the Java man and Peking man, but have since discovered similarities with East African hominids from Olduvai Gorge in Tanzania and Lake Turkana in Kenya. Chinese archeologists theorize that Wushan man may represent an earlier form of man than was previously known to exist in Asia, one that may have either developed independently from the African hominids or branched off from them at an early date. Chinese researchers have long hoped to prove that their ancestors originated in China rather than in Africa or anywhere else. Some historians in Wushan have seized upon the archeological find as evidence of this, and a local introduction to the county written in 1997 states that there is "reason to believe that the forefathers of the Chinese race, the forefathers of the yellow races, and the forefathers of other races sprang from the highlands of the Three Gorges."[1] While this idea is popular along the Yangtze River, it has yet to gain general acceptance elsewhere.

Wushan man is known to have coexisted with Gigantopithecus, the ancient ape that has been found in Sichuan, Hunan, and Yunnan provinces, and in Vietnam. An enormous slow-moving animal about six feet tall and weighing over 400 pounds, the ape became extinct somewhere near the middle Pleistocene, possibly around 200,000 years ago. One theory is that human hunting may have led to its demise. Wushan is one of the three known sites where fossil remains provide evidence that Gigantopithecus and early man were both there at the same time. Gigantopithecus was a furry animal that walked on its knuckles but looked human when it stood at full height. For as long as anyone knows, reports have persisted throughout Asia, and in Hubei and Sichuan in particular, of great half-human, half-apelike beings. Reminiscent of stories about Big Foot or the Abominable Snowman, the *yeren* (wild man) walks the hills near the Yangtze, sometimes kidnapping women, children, and animals.

Upper Paleolithic (ca. 45,000–8000 BC) and Neolithic (ca. 8000–2000 BC) remains have also been found along the upper Yangtze. These include remnants of the Daxi Neolithic culture, a rice-growing society that flourished along the banks of the Yangtze River in eastern Sichuan and western Hubei from about 5500 to 3300 BC. Archeologists have unearthed their distinctive red and black pottery, jade, stone, and bone wares from over twenty sites along the Daxi River near Fengjie. Nothing is known of why they disappeared, or who, if anyone, occupied the area immediately thereafter, though some archeologists speculate that the Ba people who later inhabited this region may have originated here. The late Neolithic period overlaps with the Xia dynasty (ca. 2200–1600 BC), the earliest of China's semihistorical dynasties, and the gods and semidivine emperors who preceded it. According to legend, probably around 2900 BC, after a period of great chaos, a god who lived for ten thousand years and grew ten feet a day separated heaven and earth. Three primeval emperors, Tian Huang, the Celestial Emperor; Di Huang, the Earthly Emperor; and Ren Huang, the Human Emperor, came next. A series of powerful emperors of heaven and earth followed, eventually leading to the Yellow Emperor, or Huang Di, who is considered the father of Chinese civilization. The beginning of his reign is estimated at about 2697 BC and marks the beginning of the often-quoted "five thousand years of Chinese civilization."

Various gods and emperors succeeded him, among them Fuxi, the oxtamer who domesticated animals, and Shennong, who invented the plow and gathered wild herbs for medicines. These early semidivine leaders put the world in order and taught people the basic skills necessary to live as civilized beings. During the reign of Yao and Shun, the fourth and fifth emperors, there were great floods. Yao put one of his officials, Gun, in charge of flood control. Despite many years of effort he failed, but his son, Yu, took over and succeeded. He then unified the territories drained by the Yellow, Han, Huai, and upper Yangtze rivers, and eventually became the first emperor of the Xia dynasty. Its founding date is put at 2205 BC, a time when history and legend are intertwined. While historians believe that Emperor Yu, if he existed, was probably born in the north of China, the mythology about his role along the Yangtze gorges is known throughout China and is thoroughly entrenched in the regional culture. Nowhere is this more so than in Wushan, where he appears battling the rising waters in countless folk tales.

Many of these stories emphasize the crucial assistance provided to him by Yaoji, the fairy who became the Goddess Peak, one of the most striking rock formations high above the Yangtze. Yaoji brings good fortune to those who glimpse her through the mist. Like the Lorelei on the Rhine, the goddess looks down upon the passing ships, but protecting those making the treacherous journey rather than distracting them. Yaoji was the twenty-third daughter of the King of the Western Heaven. Bored with life in the heavens, she studied magic and then suggested to twelve of her sisters that they come with her to take a look at the world. Flying above Wushan, they saw twelve evil dragons killing people and destroying fields and forests. (Twelve is an important number because of the lunar calendar and the signs of zodiac, but almost everything of significance in China is counted: the Three Principles of the People, the Four Old Things, the Five Poisonous Creatures, and so forth.) Yaoji quickly put a spell on the dragons and killed them, turning their corpses into large boulders, which unexpectedly dammed the Yangtze to the east of Wushan and caused a terrible flood. Yu, with his expertise in water control, arrived immediately but could not figure out what to do. Instead, he sighed so much that he raised great winds and made matters worse. Yaoji calmed him down and ordered bolts of lightning to build up the embankments along the shore and thunder to break up the rocks and release the water.

Yu went to thank her, but she was nowhere to be found. Strange objects around him kept becoming other things—first a green rock vaporized into green smoke, which turned into a dragon and then became a white flamingo. A heavenly guard finally explained that all the strange things that had appeared were actually Yaoji in different forms. Yu bowed to her, and she went back to her original, recognizable state and taught him more about taming rivers and slaying dragons. After saving the people who lived along the Three Gorges from the water, Yaoji and her sisters decide to stay there and to help with other problems. The fairies saw that harvests were bountiful, helped the sick, watched over sailors, and protected wood gatherers from wild animals. In time, Yaoji, ever watchful from the highest point of Wu Gorge, turned to stone and became the Goddess Peak. Her sisters remained with her and became the peaks of Mount Wushan.

The massive stone mountain known as Shibao Block (for which the town is named) supposedly was also formed as a result of divine intervention by fairies who coexisted with Yu and Yaoji. The tales involving mythic conflicts are generally straightforward—gods battle against nature or one another, something in nature is damaged, chaos results, and then order is restored. One version of the founding and naming of Shibao Block follows this pattern. In this, during an argument between the God of Fire and the God of Water, the God of Water accidentally knocked over the mountain that holds up the sky. As the mountain tipped, it tore a hole in the sky and fire and rain poured out and monsters and demons escaped onto earth. Nuwa, an ancient goddess who had long ago created people out of clay and later helped the Great Emperor Yu carve out the Three Gorges, awoke from a deep sleep and quickly hammered out a set of jade stones. She used most of these to plug the hole in the sky, and then set about dealing with the monsters. First she hurled her last jade stone at a demon turtle and killed it. Then she threw her stone at a nine-headed dragon, a follower of the God of Water, who had started the problem with his clumsiness. The badly wounded dragon tried to flee, but Nuwa aimed at him again. As the stone bounced to the ground, it turned into a mountain and crushed him. Meanwhile, there were other problems. The sea, which was nearby, began to overflow and seep out through the earth. As the water surged forth from the ground, the mountain stopped the flow. Thanks to the new mountain, the sky was repaired, a disastrous flood averted, and the demons obliterated. Henceforth, the local people called the mountain Shibaozhai, or Precious Stone Block, as it once was a precious jade stone. To this day, some people believe that it has special powers to protect the village and suppress evil.

Other common regional tales bring together a blend of ancient legend and continuing themes of illness, recovery, greed, and revenge. While ordinary people can recite local stories in sometimes excruciating detail, their importance or why they are so well remembered is not always clear, other than that the characters do the kind of things that real people do and sometimes suffer the same consequences. If you ask anyone on the dock at Wushan about the significance of the story of the gods who kept the flood waters at bay or of the great military strategists, they will tell you about the founding of Chinese culture. If you ask the meaning of a story of a rock, the local fishermen and the local historian will probably be puzzled by your stupidity and say, "It's a story. It explains how it got here." How it got here is of utmost importance culturally, not only in Wushan but everywhere in China, and endows everything with a past.

A different story about the naming of Shibao Block fits into this category and provides an alternate explanation for how the mountain received its name. In this account, Hexiangu, the only female of the Eight Immortals (the *Ba Xian*, a traditional grouping of Chinese deities who live in the Penglai Isles, the location of the Eastern Paradise), becomes bored with her life at home. After traveling around the countryside looking for a quiet and secluded place, she decides on Shibao Mountain, which already existed but was not yet named. Disguised as a beggar, Hexiangu goes down to explore the village where she discovers that the residents are all sick and dying because of a shortage of medicine. She learns that the pharmacy has the necessary medicine, but the evil owner will not sell it because he wants the local people to die so he can seize their lands and forests.

On the road to town, the immortal meets a little girl named Shibao whose parents have died from the epidemic. Hexiangu adopts her and then flies back to Penglai and asks the others to come help. The seven other immortals bring back medicinal herbs and plant them on the mountainside. In time, the townspeople recover and are thankful. Though the other immortals soon fly home, Hexiangu stays on to teach Shibao how to grow herbs and cure illnesses. When Hexiangu must return to Penglai, Shibao, now a beautiful young woman, devotes herself to helping the poor and sick. This makes the pharmacist even angrier because he had already lost so many of his clients and he still has none of their land. He had hoped that once Hexiangu left he would become powerful again. Day after day he broods about the trouble Shibao has caused him, and finally he climbs up the mountain to kill her. Shibao learns that he is coming and destroys the plank road on the side of the cliff, the only route to her mountain home.

Frustrated, the pharmacy owner sets fire to the mountain, but a hot stone falls and hits him on the head and kills him. As this happens, a hawk flies out from the flames toward the east. The fire burns on the mountain for seven days, but never harms the herbs. After the fire is spent, no one can find Shibao or her body. The villagers realize that the girl must have turned into a fairy and flown away in the form of a hawk to join the Eight Immortals. In order to commemorate her, the local people called the mountain Shibao Mountain and built a temple upon it.

A closer examination of such legends about fairies and gods reveals something of the time period in which they originated, and how characters that emerge in one era recur and are intermingled with those of later dates. Emperor Yu and his complicated array of ancestors are described as early as the second century BC and belong to a part of the past some four thousand years ago, once assumed to be myth, that archeological discoveries now suggest is likely to have an historical foundation. References to the ancient Sichuan sea, holes being torn in the sky, and the Eastern Paradise in the Penglai Isles likewise go as far back as the Han dynasty (206 BC–AD 220). Nuwa, who created the human race out of clay and appears again and again throughout time to assist people, is an ancient female goddess figure. Her origins are unknown but her roots are said to trace back to the ancient period of Chinese shamanism.

Hexiangu, who is not believed to be an historical figure, was according to legend born in Guangdong province in AD 700 with only six hairs on her head and never to have grown any more, though she is depicted with a full head of hair. Usually shown with a magic lotus, she became an immortal by eating a powder from the Mother of Pearl Mountains, as instructed in a dream. The Eight Immortals, in contrast, are relatively new additions to Chinese mythology. Based on historical and fictitious figures who lived at times ranging from the second to the eighth century AD, they did not appear together as a group until the Yuan dynasty (1279–1368). The juxtaposition of the girl Shibao and the spirit Hexiangu is an anachronism that probably puts the origins of this version of the story somewhere after the fourteenth century. Greed for land is age-old, but an even more contemporary contribution to the story is suggested by the lengths to which the pharmacist went to obtain other people's holdings. This has a modern ring to it, reminiscent of the evil landlords of the early twentieth century.

Ancient legends as well as archeological discoveries suggest that the Yangtze was well populated in ancient times. The area between Yichang and Yunyang, particularly near Wushan and Xiling Gorge, has yielded an array of Shang dynasty (ca. 1600–1066 BC) artifacts, among them a bronze vessel (*zun*) found on the upper Daning River in Wushan county. Similar to finds elsewhere in Sichuan, this is considered strong evidence of an active bronze-working culture stretching across Sichuan province and southern China at this time. Throughout most of the Western Zhou (ca.1066–771 BC) and Eastern Zhou (770–221 BC) dynasties, when China was divided into small kingdoms warring with one another for supremacy, the upper Yangtze was home to the Ba people, whose kingdom extended from Chongqing to near the border of Hubei province. They are believed to be the ancestors of the Tujia, an ethnic minority people who still live in this area. Little is known about the origins or fate of the Ba, a warlike people who buried their dead in boat-shaped wooden coffins wedged into the sheer cliff walls high above the river, though many artifacts of their civilization survive and can be viewed in museums in Chongqing and elsewhere in China. Sima Qian (ca. 145–85 BC), known as the grand historian of China, claims that the first Ba was a son of Fuxi, the ox-tamer, and thus a great-grandson of the Yellow Emperor.

The *Shan Hai Jing* (The Classic of Mountains and Seas), a first-century BC collection of geographic lore and fact, offers a more complicated version of the origins of the Ba people. According to it, the Ba were one of five clans who lived on a mountain in Hubei province. Women from four of the clans gave birth in a black cave, but the Ba were born in a red cave. After many years with no common leader, the clans held a competition to choose one. There were two required tasks—to throw a sword into a small hole in a rock and to make a boat out of earth that would float. A Ba clansman named Wuxiang succeeded at both and was declared the leader. He immediately sailed away in his earthen boat down the Qing River in southwestern Hubei, perhaps to escape some threat that history has forgotten. Here he met a goddess who offered to share her lands with him, but Wuxiang (who had taken the name Linjun) wanted them all for himself. The goddess tried to get rid of him by turning herself into an insect and leading an army of gnats to block the light of the sun, but Linjun waited until a hole in the swarm of bugs let through a ray of sunlight. He then shot and killed her with an arrow. With the goddess and the insects out of the way, he and his followers founded a town called Yicheng just south of the Sichuan border near the site of the present-day town of Enshi. When Linjun died, he was reincarnated as a white tiger, supposedly giving rise to a Ba ritual of giving tigers human blood to drink. From Yicheng, Linjun's descendants migrated westward, settling along the banks of the Yangtze in what is now Sichuan province.

A disconnected and dreamlike series of events, this story of the Ba is marked by violence and indifference, both common in Chinese folk tales. However, unlike similar stories that pertain to individual towns or villages and their surroundings, myths about the Ba are not well known along the river and do not generate much popular interest. In the gorges, local people point out with perplexity the holes in the cliff walls where the Ba buried their dead. The general consensus is, as a Fengjie storekeeper put it, "They were strange." The Ba are believed not to have been Han, the ethnic group to which over 90 percent of Chinese belong, and the idea of hanging coffins and tigers drinking human blood does not resonate much with the Chinese here. They are someone else's ancestors. From a local perspective, the Ba are an historical curiosity worthy of note, but few of the people who farm and fish here have any sense of identification with these ancient warriors who preceded or eluded Chinese civilization, nor with the Tujia people who still live near the Sichuan-Hubei border and are believed to be their descendants.

The main capital of the Ba kingdom was in present-day Fuling, with regional strongholds in Fengjie, Fengdu, and Badong counties. For centuries, villagers speculated about the contents and purpose of the black boxes jutting out from the sides of the cliffs. Some said they were bellows left there by Lu Ban, a legendary craftsman who came with Emperor Yu to help him dredge out the Three Gorges, a story that gave this section of the river the name of Bellows Gorge. Others speculated that the strategists of the Three Kingdoms period kept their books and papers here. In 1971, a research team reached and opened the coffins and found human bones and teeth, along with ornaments and tools.

According to historical records, the Ba were renowned for their war chants and drums. They fought frequently over territory with the armies of the Chu, an adjacent state covering all of Hubei and much of Hunan, as well as the central plains of China north of the Han River. There were intermittent alliances between the two, but in 634 BC, the Chu attacked and annihilated Kui, an independent city-state on the site of modern day Fengjie, which served as a buffer zone between the two larger kingdoms. The Chu then pushed the Ba farther westward, while the Qin, who were later to unify all of China, attacked them from across the Qinling mountains to the north. By 316 BC, the Qin had conquered and occupied most of the territory of the Ba as well as the kingdom of Shu, located to the north surrounding what is now the city of Chengdu.

Under a hundred years of Qin domination, Shu became increasingly unified, a result of the more authoritarian policies enforced there, while southern Ba retained a greater independence and cultural identity. In Shu, the Qin first implemented the innovations and improvements in infrastructure, including irrigation, road building, and land reform, that the dynasty was to institute throughout China. The most famous feat of this era was the Dujiangyan reservoir and irrigation system completed near Chengdu in 285 BC. Built by hundreds of thousands of laborers, it was the largest public works project in Asia ever constructed at that time. It remains there still, visited by thousands of tourists every year. The Qin also began the tradition of using faraway and uninviting Sichuan as a place of internal exile and dispatching thousands of people whom they found objectionable for one reason or another to the towns and cities of Shu. The exiles were free to pursue economic opportunities and often prospered, but checkpoints were rigorously enforced. Less dominated by the Qin, Ba was not considered secure enough for banished officials. Only in succeeding dynasties did the towns of the upper Yangtze become a destination for those wanted out of the way by the ruling government.

The population of the Ba region of eastern Sichuan had always consisted of a variety of ethnic and tribal groups that the Ba dominated and controlled to varying degrees at different times, and the Qin continued to recognize their primacy among other ethnic groups after 316 BC. Ba aristocrats were allowed to keep their titles, though they were expected to pay cash tribute to the Qin. Commoners were required to provide chickens and cloth. Qin officials in return sent gifts of women from the Qin empire to the upper classes. The marriages and families that resulted helped ally the leadership of the territory more closely to the Qin rulers. While the Shu periodically attempted to revolt against the Qin and the onerous conditions they imposed, which included heavy taxation, corvée labor, and forced migration to and within the region, the Ba coexisted relatively smoothly with their Qin rulers. The most significant changes at this time were economic, and it was during this period that cash replaced barter. The small number of migrants to the region, undisturbed political structures, and difficulty of transportation left Ba relatively stable, though its importance as a political entity faded over time. Continued Chu incursions into Ba territory were a significant factor leading to a loss of Ba political power, and many of the aristocratic Ba families and their armies were eventually driven northward up the Jialing River toward the Shu Kingdom and present-day Chengdu.

In time, the Chu lost out to the Qin as well. One of the most famous tragic stories of the Yangtze River arises from this period. Qu Yuan, the vice-premier of Chu who was born in Zigui, near the border of Sichuan and Hubei provinces, had advocated drastic internal reforms and advised the formation of an alliance against the increasingly powerful Qin state. His counsel was ignored, and Qin officials bribed sympathizers at court to induce the king of Chu to visit Qin, where he was captured and held until he died. Falsely blamed for this and other troubles, Qu Yuan was banished to the south of the Yangtze, in today's Hunan province. In 278 BC, the Qin captured the capital of Chu. Qu Yuan, overcome with sorrow and anger at the defeat of his country, threw himself into the Miluo River and drowned on the fifth day of the fifth month of the Chinese lunar calendar. This date has been commemorated ever since with dragon boat races, which recall the frantic search for his body, with people in boats shouting out his name and beating drums. Despite their efforts,

Qu Yuan's body was never found in the river because a fish had swallowed it and carried it back to Zigui for burial. Six years later, in 272 BC, the Qin successfully stormed the city of Wushan, the last stronghold of the Chu, after which time the Qin had unchallenged control over all of eastern Sichuan.

In 221 BC, fifty years after the fall of the Chu, the Qin united all of China, but the emperor, Qin Shi Huangdi, died in 210 BC and his dynasty (221–207 BC) survived only three more years. He had initiated massive infrastructure programs throughout China, among them the Great Wall and the Ling Canal that connected the Yangtze with the Pearl River in the south, but he forced people to work under brutal conditions. As resentment grew, corvée laborers joined the ranks of rebellious militias. Small revolts grew into large ones, and Qin Shi Huangdi's successor could not control the unrest. Several years of struggle between regional military leaders followed. In 206 BC, Liu Bang, a former Qin official who came from an area just north of Ba in the Han River valley, won out and founded the Han dynasty.

The early days of the Han dynasty (206 BC-AD 220) were a period of relative stability and growth along the upper Yangtze. Sichuan was known collectively as Yizhou, but the traditional distinction between Shu in the north and Ba in the south continued. In the census of the year AD 1, the figures for Ba record 158,643 households and 708,148 people, while the larger Shu had 268,279 registered households and 1,245,929 people.[2] Although the census was not all-inclusive, one can see that in two thousand years the population of the general area of Sichuan increased by about a hundredfold, from two million to two hundred million. By the mid-Han dynasty, Wushan and Fengjie were thriving centers of commerce. Wushan was a key transportation link between north and south, with a growing number of goods shipped to Xianyang, the Han capital in Shaanxi province, along the plank road suspended above the Daning River on beams hammered into the cliff wall. The road was also solid enough to transport the Han armies southward. Though the upper Yangtze was not frequently used for long-distance transportation of goods because of the navigational perils it posed, in 115 BC, as part of a nationwide emergency measure, tons of grain from Sichuan were moved along the river through the gorges eastward to the Huai River, where famine refugees were waiting. A series of disastrous floods along the Yellow River had left huge sections of the population of northern and eastern China starving. This was one of the most well-coordinated and extraordinary relief efforts in early China, but one not often repeated because of the overwhelming difficulty of the rapids and currents of the Yangtze, which caused many catastrophes of its own. Systematic records of natural disasters were kept by this time and show that from the second century BC to the present, the Yangtze averages a major flood every ten years and a catastrophic flood every hundred.

Many of the crops and industries that are still important along the upper Yangtze took hold in the Han dynasty. Tea production is first documented in the first century BC. Medicines indigenous to the Three Gorges region, including herbs from the mountainous area near the Shennong River, named after the ancient divinity who first identified and distributed medicinal plants, became a standard part of Han pharmaceutical resources. Salt production and iron manufacturing became major industries, and a commissioner of oranges, indicating the importance of the crop to the Han economy, was appointed in the area of what is now Wanxian. By the end of the Han dynasty, political turmoil and popular uprisings had led to a general increase in chaos and a decrease in the people's livelihoods. Patterns that were to repeat themselves into the twentieth century led to massive revolts and the eventual loss of control of the dynasty. Aristocrats and wealthy members of the upper classes kept large private armies of semidestitute peasants who eventually either rose up against them or joined other rebellions or sects that in time united in resistance against the ruling government. While beliefs about supernatural creatures go back to ancient times in China, at this time various forms of demon and ghost worship became more widespread and more systematic, and often grew into folk religions that became the focus of groups active in popular religious movements and uprisings against the government.

Popular beliefs about ghosts that linger on to the present often had their roots in the Han dynasty as well. In a regional variation of the common Chinese saying that "Suzhou and Hangzhou are heaven on earth," along the river one hears that Shibao Block is heaven and neighboring Fengdu, in the next county over, is hell. Some say Fengdu became known as the home of the King of Hell as early as the Han dynasty, but this is more likely to have occurred during the Tang dynasty, when Buddhist concepts of hell became widespread and began to merge with popular Taoist beliefs. The source of this myth seems to have been a confusion over the names of two scholars who lived on Mount Mingshan in Fengdu in the first century AD. One was named Yin, the other Wang, names which in Chinese sound like the words for "king" and "hell." Over time, the mountain became known as the home of the rulers of the underworld. By the mid-Tang dynasty, Mount Mingshan was covered with both Buddhist and Taoist temples and shrines, and the belief spread that Fengdu was a transit point for spirits of the recently deceased from all over China. Over time, the King of Hell, who was responsible for adjudicating people's fate after death, came to share Mount Mingshan with a retinue of underworld officials, including ten supreme

Figure 2.1 Demon on Mount Mingshan, Fengdu

legislators, six judicial officers, four leading judges, ten commanders of the Nether Institution, the Cardinal Ghost with the Gold Head, the Cardinal Ghost with the Silver Head, and the Cardinal Ghost with the Copper Head. After this panel came to a decision about who was destined to live again in human or other form and who was to go directly to paradise, the spirit was permitted a last look at his home and family from the Looking Home Pavilion, and then would drink a memory-eliminating soup and go on to the next life.

Mount Mingshan is now a ghoulish tourist site. Spots such as the Looking Home Pavilion, once visible only to the spirit world, have taken on a physical shape easily viewed by the living. Ten-foot statues of green and yellow demons disemboweling adults and biting off the heads of babies are easily accessible by cable car. In the town itself, a fun house brings you through the horrors of hell in a bumper car, and there is a satellite ghost museum in Wushan. Though the residents of the Yangtze towns do not take these ghosts too seriously, regarding them mostly as fairy tales, ghosts and mythical animals have remained matters of concern in the countryside up to modern times. Students sent to villages in the Three Gorges during the Cultural Revolution recall that though their mission was to destroy feudal customs and superstitions, ghosts and witches and strange immortal creatures were a source of constant anxiety during evening trips to political meetings. For the farmers as well, returning from the fields late at night was dangerous enough because of the slippery footpaths, but in addition they had to risk disappearing into the mouths of hungry ghosts, who were often spotted dangling in the trees. The spirits of men and women who had hanged themselves, known as hanging ghosts, came out after dark to look for people to eat, for only by replacing themselves could they stop wandering the earth. The captured victim was roasted and then devoured whole, leaving no trace. Killing oneself by hanging or drowning in a well has a long history in the countryside. Life here has always been difficult enough to ensure a plentiful supply of ghosts.

At the end of the Han dynasty, three small kingdoms (Shu, Wu, and Wei) emerged out of the chaos and fought for dominance. The stories of their battles and the intrigues of the Three Kingdoms period (AD 220–265) are among the most important sagas of Chinese history, strategy, and melodrama, comparable in Western culture to the stories of King Arthur and the Knights of the Round Table. History and legend are mingled almost indistinguishably in the minds of the people who live along the shores of the upper Yangtze, a phenomenon often noted by scholars and travelers along the river. As Lyman Van Slyke writes in his book *Yangtze: Nature, History, and the River*, "although at first glance the educated Chinese scholar and the illiterate tracker may seem to have inhabited different worlds, to a remarkable degree they shared 'that fund of folklore and history and myth' upon which were founded both the scholar's mastery of texts and the tracker's memory of an oral tradition."[3]

The kingdom of Shu, ruled by Liu Bei, the pretender to the throne because of his distant relationship with the Han emperor, encompassed what is now Sichuan. Liu Bei's military headquarters were in Fengjie, and nearly 1,800 years later the Three Gorges region is still dotted with memorials to him and his closest advisers. The most well known of these were Zhuge Liang, considered one of China's most important military geniuses, and Liu Bei's sworn blood brothers, the "tiger general" Zhang Fei and Guan Yu. Guan Yu, renowned for his loyalty, gradually evolved into Guan Di, the god of war, loyalty, justice, and righteousness. The Wu Kingdom, led by Sun Quan, was to the east, covering most of what is now Hubei and Hunan provinces and stretching along the Yangtze to the sea. Cao Cao, the brilliant but ruthless Duke of Wei, ruled the northern kingdom of Wei in the Yellow River basin. At the end of the Han dynasty, he appointed himself premier and was able to control the weak and ineffectual emperor. In AD 208, in an attempt to unify all of China, Cao Cao initiated a military campaign against the south, taking on both Liu Bei and Sun Quan. The outcome of the struggle for control among the three powerful leaders was tragic for all, a frequent outcome in Chinese history and perhaps one reason why the people along the river and elsewhere in China continue to feel such a close connection with these stories. The Wei defeated the Shu in AD 263, under the leadership of Cao Cao's successor, who two years later was toppled in a coup d'état from which the Western Jin dynasty (265–280) emerged. More than a decade later Jin's armies vanquished the Wu Kingdom and unified China. None of the leaders of the Three Kingdoms ever achieved his ambition to rule all of China.

Contemporary lessons in military strategy still borrow tactics from this era, and the romanticized and elaborate tales have been the source of innumerable stories, poems, operas, and books, the most famous of which is the Ming dynasty (1368–1644) vernacular novel *The Romance of the Three Kingdoms.* Mao Zedong is reported to have said that when he set out on the campaigns that brought him to power, no book was more valuable. The complicated and often confusing epic spans more than sixty years and is ultimately tragic for all the players. After a successful but short-lived alliance between Shu and Wu against Wei, Sun Quan, the leader of Wu, attempted to persuade Guan Yu to join him and betray his friend and commander, Liu Bei, the Shu ruler. When Guan Yu refused, Sun Quan cut off his head and sent it to Cao Cao, the Duke of Wei, to suggest that he might like to join the Wu or encounter the same fate.

Hearing of Guan Yu's death, Liu Bei sank into despair and began what turned out to be a disastrous campaign against the Wu. News soon reached him that his other trusted friend, Zhang Fei, had been decapitated by two of his own men who, following local custom, had intended to surrender to the Wu with their leader's head. Hearing of a temporary truce, they panicked and threw Zhang Fei's head into the river, but the general's spirit alerted a fisherman to retrieve it and bury it in his home in the kingdom of Shu. For a time, the fisherman kept the head in a vat of oil, which he made available to the lovelorn, who, for three copper coins, could ask Zhang Fei for advice. When someone posed a question, his head floated to the surface and gave advice. Eventually, the fisherman (presumably considerably richer by then) buried the head on Flying Phoenix Hill near Yunyang, the next county over from Fengjie. A temple commemorating Zhang Fei was built in the early Northern Song dynasty (960–1127). His spirit is said still to assist passing boats from the nearby Helping Wind Pavilion.

Local histories have long claimed that when the Shu troops retreated to Fengjie, Zhuge Liang built a fortress in the river, known as the *shui bazhentu*. The remaining stones of this structure caused a shoal a mile and a half from Fengjie that was a hazard to navigation until the stones were removed in 1964. The fortress, whose remnants emerged during the dry season, was made of 564 blocks arranged in a square near the shore, surrounded by twenty-four stone pillars, each five feet high and six feet wide. Despite the extraordinary construction, the Wu fleet managed to break through and defeat the Shu. Two centuries later, in his poem "Bazhentu," the poet Du Fu wrote about Zhuge Liang, "He earned everlasting fame for his stone fortress. The river flows on, but his achievements still stand. It is regrettable that his blockade failed to engulf the Wu troops."[4] An unexpected twist to this story was that when the stones were removed in the 1960s, archeologists found that they were not from the Three Kingdoms period at all, but date from a far earlier time, probably the New Stone Age (ca.8000–2000 BC) and the Bronze Age (ca. 2000–1100 BC). They are believed to have been used for ceremonial and sacrificial purposes.

Liu Bei's army trained in what is now downtown Fengjie. After being defeated by the Wu army, Liu Bei died of despair in his Eternal Peace Palace, where the Fengjie Teacher Training Institute is now located. The school has built a small garden in his memory, which contains ancient tablets commemorating his death. According to other versions, Liu Bei actually died on Baidicheng, where a pavilion with replicas of the heroes of the Three Kingdoms now stands. Two halls contain excellent examples of Song and Ming stele; outside is a garden in which stone engravings of modern love poetry have been added. Nearby is a thatched cottage, many times rebuilt, where the poet Du Fu is said to have written when he was serving as an official in Fengjie.

Up and down the river, memorials and inscriptions invoke the struggles and tragedies of the Three Kingdoms period. On the Kong Ming Tablet (Kong Ming is another name for Zhuge Liang), an oblong rock in Wu Gorge, are the words "Wu Gorge has peaks rising higher and higher," reputedly written by Zhuge Liang, but actually inscribed during the Ming dynasty as a tribute to him, along with comments on the unsuccessful Shu-Wu alliance. Farther downstream in Xiling Gorge is the "Gorge of the Sword and Book upon the Art of War." On the cliff stands a rock resembling a book, out of which projects a swordlike stone. In the days before his fame, Zhuge Liang had written a book on military command, but sick and unable to find anyone interested in it, he put it away on the cliff in the hopes that someday an appreciative reader would come upon it. All of these, as well as innumerable other historical and archeological sites, will be transplanted, rebuilt, or lost once the floodwaters from the Three Gorges Dam have risen.

Of the approximately 1,300 known archeological sites in the area to be flooded, archeologists have determined that between 400 and 500 are worth saving. Other cultural legacies, such as the "low-water calligraphy"—the poems and other writings engraved on the walls of the upper Yangtze, some of which indicate safe passage levels for boats and record low water levels—will be submerged under hundreds of feet of water. The oldest known engraving along the river, a Tang dynasty (AD 618–907) depiction of two fish whose eyes appear to be open when the water is deep enough for navigation, is part of the Baiheliang engravings near Fuling that have been declared a national-level monument. These consist of over 30,000 characters in inscriptions from the Tang to the Qing dynasties. In anticipation of the great increase in water depth as a result of the Three Gorges Dam, attempts are under way to carve out a portion of the inscriptions, but the limestone walls of the river and underwater conditions make this difficult to accomplish, and only a small number of engravings will be preserved.

Each county has its own Cultural and Antiquities Bureau that is responsible for local record keeping and access to monuments and historical sites. These have more recently become the local liaisons with archeologists from Beijing and elsewhere who are cataloguing and preserving items and sites of importance. Since 1994, the Museum of Chinese History in Beijing and the State Bureau of Cultural Antiquities have been designated as the units responsible for the preservation and protection of cultural sites in and around the Three Gorges. Twenty-eight other academic institutions are also involved in the project. Not surprisingly, the perception of what are acceptable cultural losses varies greatly among the people who live and work here, archeology and conservation experts, and the members of the bureaucracy dealing with these issues. For many of the archeologists and art historians, this is a national disaster in the making, a loss not only of what is known to be here now but also of future discoveries. This is particularly true of archeological sites, many of which were first identified and excavated in the 1980s and 1990s. For the planners of the dam, money and the rapid completion of the project are balanced against cultural loss. Inevitably, the dam wins out.

CHAPTER 3

TRADITIONAL LIFE

SINCE THE BEGINNING OF TIME, THE STORY of the Three Gorges region has been one of the violence of nature topped off by the chaos of people. The customs and traditions of the region, some of which have lingered on to the present, developed to a large extent to cope with this. The compilations of historical facts issued every five years by county governments throughout China tell of 2,000 years of constant struggle against water and heat, of crops rotting, of people and animals being washed away in rain and floods, and of fields shriveling and cracking in droughts, followed by famine when there were no harvests. The Wushan and Fengjie county archives record thousands of big, great, and disastrous droughts, floods, destructive hailstorms, ice storms, fires, and earthquakes between 190 BC and the year 2000. A sample of local tribulations includes a great flood in 190 BC, a big drought in AD 647, and another big drought and famine in 1058 encompassing both Wushan and Fengjie counties. In 1340 it was dry, in 1570 there was a disastrous drought and famine, and in 1684 all the fields dried up. By the first half of the nineteenth century, the cycle of drought and flood had speeded up to the point where famines occurred at least once a decade. Some years receive a special note because starvation was so widespread among animals as well as people that wolves came into the city and ate the townspeople. The same listings of various forms of misery repeat themselves in county after county.

From local histories, poetry, travelogues, and the half-remembered stories of the parents and grandparents of people now living, we can glimpse the daily life of the past and its connections with the present. Rural revolutionaries worried about ghosts in the 1960s because of the ghosts of the past, and peasants bury their dead with footrests embroidered with bats for good luck, and fishermen never turn a fish over on a plate lest the boat go over too, all because of traditions that have grown up since ancient times. During the thousand or so years between the end of the Tang dynasty in the tenth century and the mid-nineteenth century, life along the upper Yangtze continued with remarkably little change. It was during these centuries that many distinct customs and practices, some still familiar, took hold here and a strong local identity based on isolation and independence was consolidated.

The present-day towns in the Three Gorges were already well established by the Tang dynasty, though in a few cases their names have changed. Fengjie has moved back and forth across the Yangtze five times since the eighth century BC and found its most recent resting place on the north bank of the Yangtze during the Song dynasty. With the construction of the Three Gorges Reservoir, the city will be divided among three sites on both sides of the river. Other towns have also moved across the river or slightly up or downstream, usually as a result of natural disasters or warfare that destroyed or decimated them. Some communities have prospered in comparison with the past, just as many have declined, having lost whatever claim to economic or cultural vibrancy they once had. Daily life did not change much from century to century. For most people, the days were governed by the seasons and by the cycle of years when crops were plentiful, followed by years of flood and drought. Village and town life was punctuated by birth, marriage, and death, and the celebrations and rituals that accompanied them. Weddings usually involved a series of Confucian-based rituals honoring the couple's parents, while funerals incorporated a more lively combination of Taoist and Buddhist practices. Some funeral rituals common in the Three Gorges as late as the nineteenth and early twentieth centuries, and still occasionally seen today, are thought to have been derived from the ancient musical traditions of the Ba people 1,500 years earlier. In Wushan and Zigui, near the Hubei border, villagers observe the death of family members by chanting and dancing to the rhythm of a solitary drum. This mourning dance was believed to reduce the tally of sins committed by the deceased and to help the soul of the deceased rise to heaven.

One of the aberrations of mid- and late-twentieth-century China was the enduring desolation of the towns and cities that followed after the obliteration of most traditional forms of entertainment and relaxation during the Cultural Revolution of the 1960s. Some of the old festivals were revived in the late 1980s, and in the county seats there are tea houses, restaurants, modern bowling halls, and movie theaters. The streets of many small towns, backwaters

Figure 3.1 Threshing wheat in the streets of Dachang

now as in the past, again offer at least as much to enliven the senses as they did 700 or 800 years ago. Celebrations held according to the Chinese lunar calendar, such as New Year's and the Harvest Festival, broke the monotony of the year in the countryside before the 1960s and do so again now. In the larger towns, there were temple fairs and night markets with storytellers, jugglers, astrologers, musicians, and performances of all sorts. In prosperous times, these frequent and colorful gatherings provided an opportunity to exchange news and to attend to practical business such as matchmaking and financial transactions.

One of many lunar agricultural festivals celebrated in the Three Gorges was Li Chun, which falls in the early spring. Farmers painted their oxen or water buffalo in bright colors and dressed up as the oxen god, who both assisted with farming and protected from flood. Townspeople decorated the roads with banners, and peasants strung soybean necklaces around the necks and horns of their oxen and brought them into town for children to crawl around their legs and bellies to protect them from smallpox. The similarity of cowpox to smallpox led to a popular belief that this close contact might persuade the oxen god to keep children safe from this terrible illness. (Centuries later, in 1796, the first smallpox vaccine was in fact created in England from cowpox.) Along the Daning River, which flows into the Yangtze at Wushan, the Festival of Cheng Wanghui (the Guardian of the Land) once drew thousands of visitors to the old city of Dachang from the surrounding towns and villages every year. During the festival, which was much like a Western carnival celebration, people dressed as ghosts and monsters paraded through the town to frighten away evil spirits and then ate and drank through the night. The Japanese bombed the great temple to the Guardian of the Land in 1939, but it was rebuilt and the festival was celebrated extravagantly during the war years, a tradition abolished once the People's Republic of China was founded in 1949.

Many celebrations venerated dead ancestors and provided a link between past and present, the living and the dead, a constant theme in Chinese literature and life. Qing Ming, a day still celebrated by most of the Chinese world, is the traditional grave-sweeping day when people tend the burial plots of their family members and leave coins and paper money to help their relatives in the afterworld. In the Three Gorges, it was customary on this day for a life-size figure of a demon with a bull's head and a horse's face to wander around the streets. Normally anyone who glimpsed this creature could expect to meet his death soon, but on this day adults burned incense to ask for his protection, and children, like miniature convicts, donned paper stocks to show that they regretted their mistakes. A holiday with a similar purpose was the Feast of the Hungry Ghosts on the fifteenth day of the seventh month when the gates of purgatory were opened and spirits returned to earth to eat food laid out for them by the living. This is still widely celebrated in Hong Kong and Taiwan. All along the gorges, floating lanterns in the shape of lotus flowers were lighted for the souls of the drowned to illuminate their way back to the next world. When these tiny boats, each just large enough to carry one ghost, rushed downstream into the rapids, no one but priests dared board a junk or sail on the river for fear of disturbing this fleet of the dead.

Another important local event was Goddess Day when the fairy Yaoji, who became the Goddess Peak, was venerated in Wushan by women bringing offerings of candles and cloth shoes to her shrine. Lu You, a Song dynasty poet and official assigned to be deputy prefect of Kuizhou (now Fengjie), wrote of his visit to the Temple of the Goddess in 1169 (the translation refers to Wushan as "Shaman Mountain," its literal meaning):

> We passed the Taoist Temple of Concentrated Truth on Shaman Mountain. I paid my respects at the Shrine of the Immortal of Wondrous Works, popularly known as the Goddess of Shaman Mountain. The shrine faces directly toward Shaman Mountain, whose peaks soar upward to the sky and whose feet stick straight out into the river. Authorities on such matters say that neither Mount T'ai or Mount Hua, Mount Heng or Mount Lu can match its wonders. Of its twelve peaks, however, not all are visible. Among the eight or nine which one can see, the goddess's peak is the most slender and superb of all, a fitting place for an immortal to dwell. The priest of the shrine says that each year on the fifteenth day of the eighth month, when the moon is full, the sound of flutes and strings can be heard echoing back and forth on the peak. The mountain monkeys all begin to wail, and only at dawn does the sound gradually fade away.[1]

As elsewhere in China, the social structure of the river towns was deeply rooted in the traditional Confucian hierarchy, though society was not so sharply divided in small towns where the poor and rich were likely to have known one another since childhood. Scholar-officials ranked above all others, then came the peasants who tilled the land, then the workers, and at the bottom, the merchants. The Chinese proverb "You do not use good metal to make a nail, you do not use a good man to make a soldier" reflects the even lower status of the military. Most people were born and died in the same place and spent their lives farming as had their ancestors before them, but fortunes still rose and fell. It was not unheard of for young boys from farming families to study in clan or temple schools or for wealthier relatives to sponsor further studies for a promising student.

The imperial examinations, which existed in one form or another from AD 606 until they were abolished in 1905, made social mobility possible. Every generation had a few men who began life as peasants, succeeded at the examinations, and went on to high positions as officials and members of the gentry class. Others, from all backgrounds, obtained a basic education, but failed or went no further, and then taught in village schools or went into occupations requiring some degree of literacy. Tan Shihua, who grew up near Qutang Gorge in the early twentieth century, described the age-old social structure built around the examinations and the *Shihs*, people who passed the examinations.

> The ruling class of all China were the *Shihs*, the learned ones. Only after passing examinations and becoming a Shih, could you receive and put on a *Shen*—the formal belt of an official, a tight band inlaid with precious stones and tiny mirrors. Education meant government position and all its advantages. Shen Shihs were the nobles who possessed the land, and who could send their children to study in order that they, in their turn, could obtain government posts and, with the profits of those posts, increase their land.[2]

If the sons of a successful family could not pass the examinations, eventually their land and power would be lost also.

The archivists in every county, as mandated since the early Han dynasty, have for over 2,000 years recorded huge amounts of material and ever-expanding lists of events deemed important by the historians of the time. The documents give a sense of what was and is considered essential factual and social information. Not only are the invasions and disasters noted, but also details about local governments, the establishment of schools, numbers of animals, kinds of plants, and the names of people who achieved social recognition. For men, this usually meant passing the imperial examinations. Women, rarely recognized in life for anything, are present mainly in the long lists of "virtuous widows," women honored by the county government because they remained chaste and did not remarry after their husbands' deaths.

During the Five Dynasties (AD 581–618) and other historical periods between the end of the Tang dynasty in AD 907 and the fall of the Qing in 1911, journals, poetry, and paintings, as well as statistical information and census reports, offer some clues about life at this time, though it is far from a detailed picture. The rise and fall of various dynasties had relatively little long-term impact on most people in the Three Gorges, with the exception of officials loyal to an era coming to its end or communities caught up in the violence that accompanied political change. With each change of dynasty, there was localized resistance and then defeat or acquiescence, after which life went on much as before. Wherever the new capital was, it was far enough away from the upper Yangtze to make it difficult for any regime to govern too closely. Regardless of what was happening in the rest of China, local struggles for power were more important and the main reason for persistent regional warfare. Ingenious contraptions, such as the great iron chains that could be raised and lowered across the river near Fengjie and in Xiling Gorge, entrapped marauding ships as early as the tenth century and were frequently used during the constant local battles.

Despite the wear and tear of warfare and the battering of nature, the region nonetheless had long periods of stability. In the Song (960–1279) and Yuan (1279–1368) dynasties, local crops, mainly potatoes, millet, taro, barley, and corn, all still grown here, adequately supported most of the relatively small rural population. The few good-sized cities had functioning economies based on the salt trade and the growing export of luxury goods such as tea and silk from the interior to other parts of China. Though scarcely rated highly as desirable posts for officials from other parts of China, Kuizhou (Fengjie), Zhong (Zhongxian), and some of the other upriver county seats in Sichuan were important regionally and were sufficiently well off that magistrates assigned here could extract an adequate income from the residents. The area near Xiling Gorge in today's Hubei province was far poorer and more backward. The county seat of Badong, located between Wu and Xiling Gorges, was sparsely populated, yet as early as the Song dynasty people lived on the edge of starvation. A tenth-century inscription attributed to Kou Zhun, a Northern Song statesman who served as magistrate of Badong at the age of twenty and later rose to the rank of prime minister, says of the state of affairs here, "It is deplorable that there have been few good officials since ancient times; it must be known that there are many suffering people in this world."[3] He is remembered for introducing what were then modern methods of agriculture and for his vigorous campaign against corruption. According to local records, when Kou Zhun left Badong in 980, the town had "neither barren lands nor vagrants."[4]

In the eleventh century, two contemporary Song dynasty officials, Fan Chengda and Lu You, left detailed records of their trips through the gorges to take up new postings, Fan in *Notes on Travel by Boat* and Lu in *Journey into Sichuan.* Both marveled over the scenery, complained about the inconvenience of travel, and described their visits to tourist sites such as the Three Travelers Cave and Baidicheng, which are still on every traveler's agenda. Fan Chengda described the method of agriculture practiced in parts of the Three Gorges at that time, a modified slash and burn system that was considered backward compared with other places even then.

> Yu fields are the lands in the Gorges region cultivated by felling and burning the natural vegetation. With the arrival of spring, [the farmers] first make cuttings in the hills, clearing away all of the forest. When the time comes to plant, they await the onset of the rains, then in one night burn the fallen trees to use the ash as fertilizer. The following day the rains come: and taking advantage of the prepared soil, they sow the seeds. In this manner the sprouts grow luxuriantly and the farmers double their harvest; but without rain, nothing of the sort will occur. The mountains are mostly covered with stony, barren soils, low in fertility, which must be repeatedly treated by the fire-field technique before planting can begin. In the spring they sow wheat and beans, from which they make cakes and dumplings to tide them over the summer season. . . . Although they may pass their entire lives without tasting rice, they have never suffered from hunger.[5]

Though this method produced a sufficient amount of food in the heavily forested regions of Qutang and Wu Gorges, it did not adequately sustain even the small population in Xiling Gorge with its rocky and infertile soil. Two hundred years after the departure of the reformer Kou Zhun from Badong, hardship again befell the county. Agriculture failed and corruption thrived.

Lu You was not impressed. He wrote, "The town is unimaginably bleak and desolate. There are hardly more than a hundred houses, and from the magistrate's office on down, every building has a thatched roof—there's not a trace of tile."[6] Lu reported that official posts in the gorges were often vacant for years because officials refused to serve in the impoverished region, and when they did, they extorted every cent from the struggling population, a situation that was becoming increasingly common as it became an institutionalized practice. A thousand years later, Badong, like many other cities in the gorges, still struggles with unemployment and corruption. One of the first cities to vacate its apartment buildings along the riverbank in preparation for the new dam, it quickly took on the ghostly look of a semi-abandoned inner-city neighborhood transplanted to the edge of the Yangtze. Shattered glass clung tentatively from the rows of broken windows, and the occasional silhouette of someone moving about in an isolated lighted apartment was visible to passing ships.

During the Ming dynasty (1368–1644) life in the Three Gorges, as well as in much of the rest of Sichuan and Hubei, took a turn for the worse. Many of the city walls along the Yangtze were built or reinforced in the early Ming, reflecting a newly perceived need to protect the communities in a more organized manner from external threats. The city walls were forbidding and impressive, with flags waving over the city gates and guards overlooking the river for signs of unfriendly ships in an increasingly turbulent society. Until the Ming dynasty, the population in the region had grown at a steady but controlled rate, and the needs of the people and natural resources were more or less in balance. In the fourteenth and fifteenth centuries, the population shot up suddenly and then decreased even more precipitously toward the end of the seventeenth century as a result of warfare and famine. Epidemics ravaged China as well, and descriptions of illness suggest that some form of plague was widespread at this time. By the beginning of the Qing dynasty (1644–1911), the population of many of the counties in the Three Gorges had decreased by as much as 90 percent.

The first half of the seventeenth century was a period of great instability, in part a result of unusually severe weather, particularly between 1626 and 1640. The droughts and failing harvests led to a sudden increase in the number of roving bandits who went from place to place terrorizing the population in search of food and goods. Reports of famine and cannibalism were frequent. Peng Zunsi in *Shu bi* (a Qing dynasty history) describes this horrible situation.

> At the end of the Ming dynasty, people ate people. First there were two years of drought, and vast farmland became barren land. A *dou* (equivalent to ten liters) of unpolished rice was worth twenty taels of gold, a *dou* of wheat was worth seven to eight taels of gold, but there was nothing to buy. People began to eat leaves. It is said that there were people with bowls of pearls who could not exchange them for a bowl of noodles. Others with hundreds of taels of gold died of hunger because there was no food to buy. Therefore people ate each other. Killing took place between fathers and sons, elder brothers and younger brothers, husbands and wives. Dead bodies could be found everywhere.[7]

In the 1640s, rebel troops belonging to Li Zicheng, who was to bring down the Ming dynasty, and Zhang Xianzhong made two forays into the Three Gorges region. Regional histories record the extensive murder and destruction of property left in their wake. In 1644, the first year of the Qing dynasty, Zhang Xianzhong and his troops stopped in Wanxian for three months on their way to Chongqing, where they remained briefly before going on to Chengdu to establish the short-lived new Great Western Kingdom. Local stories say that he was a Dracula-like figure who killed as a hobby, not content unless he murdered at least a few people a day. One night, finding no one due to be executed, Zhang murdered all his wives and concubines. The next morning when he awoke, he called out for them, but there was no response. When he realized what he had done, he blamed his aides for not stopping him and then killed them all and his slaves as well. After three years of barbarous rule in Chengdu, he was killed in 1647 by the Manchus, the rulers of the Qing dynasty.

Sichuan continued to be ravaged by war and other disasters during the early years of the Qing dynasty as well. From 1673 to 1683, almost all of south and southwest China was affected by a rebellion, known as the Revolt of the Three Feudatories, against the new dynasty. This devastated Sichuan's economy and seriously disrupted life in the Three Gorges. Although Wu Sangui, a Chinese general named as a Manchu prince, succeeded in ousting the last claimant to the Ming throne from southwest China in 1662, the area did not come under the direct rule of the Qing government in Peking but remained under the control of Wu and two other generals who had taken part in the battles between the Ming and Qing. Each general held vast territories, comparable in total size to all of France and Spain, which they successfully ran as independent kingdoms. When it became clear that the Qing Kangxi emperor did not intend to allow their rule to become hereditary, Wu Sangui, who occupied southern Sichuan and Hunan, Yunnan, and Guizhou, renounced his loyalty to the Qing and declared the establishment of a new dynasty. The two other generals, Geng Jingzhong and Shang Zhixin, joined him in support. Over the next five years, there was constant fighting in the gorges, with great harm inflicted on Wushan, Fengjie, Wanxian, and Yichang. The Three Feudatories were defeated after months of bloody warfare in January 1681, when the Qing army occupied Wanxian. Tens of thousands of people fled the region or were killed.

In a society so closely tied to nature and the forces of the environment, particularly at a time when disease and disaster were rampant, phenomena that cannot be explained are of great concern for what they may portend. During the Ming and Qing dynasties, local archives document numerous "peculiar incidents" without explanation or commentary. Most of the strange incidents, such as the repeated sightings of phantom soldiers, coincided with agricultural disasters and widespread rebellion. Many of these are clearly natural events, such as the shooting stars mentioned at regular intervals, but others are harder to fathom. In 1649, for example, on the eve of Frost Descent Day, the first day that frost is predicted according to the lunar calendar, a dragon appeared over Wushan, illuminating the sky and then vanishing. Another listing from the same year describes a termite-infested wood-beam house. As the termites gnawed away, the falling slivers of wood turned into copper coins which dropped to the ground and became strange plants that spread far and wide and could not be eradicated. What these plants did and why it was deemed worthy of note to the local historian is never explained.

Visions of ghost soldiers and horses are repeated throughout Wushan's history, particularly during times of military instability when the population was on edge. In 1633 and 1644, villagers in Zhaoyang Village near Wushan saw horses and soldiers on moonlit nights and heard the pounding of drums. Other reports describe clouds turning into animals. In 1662, three groups of white clouds suddenly became white horses, which then turned into a herd of a thousand sheep that walked off toward the west. A more ominous sign appeared in 1677 in the county town of Kaiping, when blood rained down from the sky and stuck like paste on the streets.

For as long as there have been disasters, or blood has rained down from the sky, gods or ghosts or monsters have been there too. Such beings have provided some explanation for the disasters and tragedies that have plagued the people's lives. Their existence also offers hope, for if they can be placated, some control over the environment may be possible. If a river god is appeased, if Guan Yu looks down from his temple, if the Kitchen God is not annoyed, life may go more smoothly in a place where the level of the river and the amount of rain determine one's fate. To deal with these and other problems, men and women prayed at the hundreds of temples and shrines scattered throughout the mountains of the Three Gorges. There were Buddhist temples with round sculptures of the Buddha and Guan Yin, the Goddess of Mercy, where women asked for a son or for the recovery of an ailing family member, and Taoist temples where priests performed magical rites and helped in the transition to the next world. The towns and villages had dozens of small and large temples for local gods, like the Sprouting God, who encouraged plants to grow in the spring, or the House Gate God, who oversaw the well-being of the household. Throughout China, the Kitchen God was by far the most important domestic god and, in the Three Gorges at least, a number of things were known to irritate him. Steaming snake or dog meat on the stove could bring on his wrath, as might hitting a pot with a bamboo brush while you were washing it, stepping on the fireplace, or leaving things on the stove overnight.

Most people paid attention to as many deities and spirits as possible, taking care not to offend or overlook any of them. Many women ran households similar to the one described in this excerpt from *Recollections of West Hunan:*

> On the altar in the middle of the hall were offerings to Heaven and Earth and the ancestral tablets. They also sacrificed there to the Year God and the Tutelary God. In the kitchen was the God of the Hearth, while the pigsty, cowshed and barns had their different deities too. Every morning and evening without fail the old lady washed her hands to bow before them with lighted incense. On the first and fifteenth of every month, she fasted to express her gratitude and pray that no harm would come to the family or to their livestock. During the festivals in different seasons she observed the appropriate rites to pay homage to the spirits, fasted to purify her mind or killed a pig to redeem a vow, unquestioningly following all the old customs.[8]

Over hundreds of years, both on the river itself and in the mountainous villages a few miles inland, numerous superstitions and traditions evolved, some unique to the region, others shared throughout much of China. Most provide common-sense guidelines about how to deal with the world of the supernatural, protect your children, and in general keep things in equilibrium. Ghosts and grumpy gods have always been a source of trouble, but while gods can frequently be reasoned with or pleased, ghosts, once mortal and longing to be so again, are motivated by envy and hunger, desires that often cannot be satisfied without harm to living beings. Customs developed to respond to the varying behavior and moods of different supernatural creatures. For example, you should not swear if you are startled by a ghoul or a ghost lest you offend it, and if you are holding a banquet, it is wise to set up a private ghost room, in case any are present. A sensible host will encourage a ghost to enjoy the meal and then ask it for assistance with financial matters. One should not speak of ghosts or supernatural beasts in the morning, particularly on New Year's Day, for this may bring trouble for the entire year. People who carry raw meat around at night should be particularly careful as they are in danger of being eaten by a ghost, along with the meat. Local people today say that the emphasis on ghosts and eating is part of an overall Chinese cultural focus on food that stems from the many years of famine. Many common superstitions are also centered on the well-being of children. If young boys or girls eat pigs' feet, it will be difficult to find suitable husbands or wives for them; if they eat chicken feet or play with magpies, their writing will look like chicken scratch. If a pregnant woman is in the house, one must not fill holes with mud or put nails in the wall lest the child be born with defects.

Other superstitions and traditions are related specifically to the river and the people who made their living on the boats and the demons who made the river turbulent and the gorges impassable. Throughout this section of the Yangtze, the rapids swallowed boats and drowned men. The worst of the

Figure 3.2 Man sleeping in boat on the Daning River

rapids were in Xiling Gorge, where the Xintan, Kongling, and Yachahe shoals follow one upon the other, but each gorge had its own terrors. Within the Three Gorges, there were seventy-two high- and low-water shoals, treacherous at different seasons, and no time of year when you could expect a safe passage. When the river was too high or low, junks unloaded their cargo, and often their passengers as well, to minimize losses. The White Bone Pagoda, with its great pile of bleached bones, stood on the north bank of the Xintan rapids, the most dangerous of the low-water shoals. This marked the burying ground of the thousands of boatmen who had drowned here. For centuries, porters and passengers climbing above the rapids passed this site, a reminder of all those who had lost their lives below.

At the beginning of Xiling Gorge, just before the Xintan rapids, is Yellow Ox Gorge, said by early twentieth-century writers to resemble the White Cliffs of Dover. The Yellow Ox, after which the gorge is named, was a local god who turned himself into an ox in order to gore through a mountain to assist Emperor Yu in creating the Three Gorges to divert the floodwaters. After completing this task, he became a 3,000-foot cliff. According to legend, the whirlpools here were made by the giant tortoise that ferried the monk Xuanzang across the Yangtze on his way to India to bring back copies of Buddhist scriptures in the fifth century AD. Xuanzang promised to find out when the tortoise would become immortal but forgot, and when the tortoise learned this (while carrying the monk across the river on his return), it was so angry that it rolled over and over, creating huge whirlpools that lasted forever. Xuanzang was saved and dried out his scriptures on the riverbank in a spot still known as the Ground for Sunning Buddhist Scriptures. Not far away, merchants and boatmen heading upstream made offerings at the Temple of the Yellow Ox for a safe trip through the gorges.

In 1854, after seeing so many boats lost in the rapids, a wealthy merchant named Li Yonggui, who lived near Xintan, raised money from traders and junk owners to build three lifesaving boats. These were placed at the rapids to assist when offerings to the gods were not enough, as was often the case. Painted bright red, they were known simply as the *hong chuan*, or Red Boats. The first three were long remembered as the *gan si dang*, the dare to die group, because of the great risks they took in saving lives. Though Li, the original benefactor, was bankrupted during the record flood of 1870, the boats were supported by other wealthy donors and officials, among them Li Hongzhang, one of China's great nineteenth-century statesmen, who after a trip through the gorges had three more boats built with subscriptions from the tax and salt bureaus in Yichang. By the turn of the century, nearly fifty Red Boats were rotated among the most dangerous spots in the Three Gorges. In 1899 Red Boats saved 1,473 lives from forty-nine wrecked junks, and in 1900, 1,235 lives from thirty-seven boats.[9]

Over time, the men on the Red Boats came to be the record keepers of the rapids, counting the number of wrecks and deaths and overseeing salvage operations. They also gradually took on the roles of river police and welfare society, buying plots to bury bodies found in the river, building shelters on its banks for the rescued, and preventing the looting of sunken ships. For their efforts, each Red Boat crew member initially received 1,200 taels of silver for each life saved and 400 taels per body recovered. Later, as it became clear that many rescue workers were reluctant to assume the burden of recovering bodies, thus taking on the responsibility for burial if family members could not be located immediately, the payment system was reversed. Red Boat workers were then paid an average of 400 taels per life rescued, and 800 taels per corpse, which allowed the rescuer to cover basic financial expenses yet still come out ahead of what he would have made had he put his efforts only into saving the living.[10] Cargo from damaged or destroyed junks was brought to nearby temples and returned to its owners or distributed to fishermen who had assisted in the rescue. This system worked well and lasted until the 1930s, when it collapsed, along with many other social institutions, in the chaos of the time.

As early as the Tang dynasty, land routes and trackers' trails skirted many of the worst rapids, and passengers routinely disembarked for foot, pony, or sedan-chair journeys of up to several days to avoid the dangers of a water crossing. In 1888, a wide path, thirty-five miles in length, was cut into the cliffs above the Three Gorges and improved transportation considerably. The trail began in Qutang Gorge near the thatched cottage of the Tang dynasty poet Du Fu, ended six miles away near Wushan county, then picked up again in the county seat and ran for another eighteen miles to the border of Sichuan and Hubei provinces. The remaining sections were in Xiling Gorge. The path, a marvel of modern construction, had stone and iron railings along the entire length to prevent travelers from accidentally falling into the river below and was wide enough for sedan-chair carriers to make their way without difficulty. In the early twentieth century, with the sudden increase in long-distance passenger traffic on both small boats and steamers, coolies stood ready at the rapids to carry suitcases to the next embarkation point. In the early 1920s, Cornell Plant, an important figure in the introduction of the steamboat to China, offered advice on avoiding the final stretch of the Xintan rapids. He wrote:

> A faulty hawser, a slippery hitch, or a false manoeuvre may mean finishing up on some jagged reef, an event that seems invariably lying in wait for the traveller on

> the upper Yangtze. There is, however, no need to take such risks. One can safely and comfortably get across the rapids by the over-land route, which lies on the left bank, i.e., the right hand side looking upstream. The landing is at Yao-tse-ngi, Sparrow-hawk Cliff, immediately above the San-tan, or Third Rapid, and close by the Temple of the Double Dragon. Here coolies may always be engaged to carry one's baggage for the modest sum of 80 to 90 cash per man-load. The pathway along the rocky bank is quite good and affords several points of vantage, when from [*sic*] one may view or photograph the rapids.[11]

Local people still use sections of these and other old tow paths as a convenient way to cover short distances, but except for a few parts that have been renovated for tourism, they are dangerous and in disrepair, and once the dam is completed, will be several hundred feet under the water. Alternative paths from village to village have long existed and the new county roads are quicker, but none offer that sense of vastness and stillness one once had looking down across the river from the crumbling stone path, where the only sounds were the wind against the cliff or the shouts of children on the rocks below and the low rumble of faraway boats.

The Xintan or "new" rapids, once the most feared of the low-water shoals, have existed since the Han dynasty, when they were created by two immense landslides, one in AD 100 and another in AD 377. Huge boulders tumbled down causing sixty- to eighty-foot waves that inundated villages as far as 30 miles away, as if they had been struck by a giant tidal wave. Until the 1950s, when the rapids were blasted away, scores of odd-shaped rocks stuck out of the water like daggers. A huge boulder, the Memba Rock, covered almost half the channel. In the low-water season, during late fall and winter when the river flows at its fastest, the channels around the rock led to a sudden drop. Ancient people said of this spot that "the water falls a hundred feet to create ten thousand heaps of snow below and a deafening noise that shakes the sky."[12] The Kongling shoal, sometimes called the Emptied Boat shoal, was the most dangerous high-water rapid until it was dynamited in 1959. Frequently compared to the Gate of Hell, the rapid was due to a 220-meter-long rock that lay "submerged in the middle of the river like a shark" and divided the river into two channels. One stream had hundreds of violent whirlpools, the other was filled with hidden rocks. Seventeen major steamship accidents occurred at this spot between 1898 and 1945.[13]

The rapids of the Three Gorges have some of the most dangerous river currents in the world, yet they have been continuously navigated with the assistance of trackers since at least the sixth century AD. Shang dynasty oracle bones depict boats in pictographs, but despite the presence of junks in Han dynasty reliefs and in the Tang murals in the caves of Dunhuang, relatively few images of boats survive in ancient Chinese art. Beginning in the Song dynasty, with the development of classical Chinese painting, with its delicate but detailed and realistic renditions of the landscape, portrayals of life along the Yangtze became a much more common subject. The detailed images of boats and trackers in the Three Gorges section of the eleventh-century scroll *The Ten Thousand Li Yangtze River* by Xia Gui, one of the most important painters of the era, offer a timeless glimpse of the river. The original scroll has been lost, but what is believed to be a sixteenth-century copy remains in the Palace Museum in Taipei.

Though river transportation and the development of canals and waterways were of chief importance in a country where the terrain often made overland travel almost impossible, the Chinese also made early forays into shipbuilding for oceans and lakes. One of the few personal records we have of early Chinese maritime exploits comes from Faxian, a monk who sailed to India and Ceylon in the fifth century AD. He kept a detailed journal, suggesting no love of ocean travel. "The great ocean spreads out, a boundless expanse. There is no knowing east or west; only by observing the sun, moon, and stars was it possible to go forward. . . . In the darkness of the night, only the great waves were to be seen . . . with huge turtles and other monsters of the deep all about. The merchants were full of terror, not knowing where they were going."[14]

As early as the eighth century, the Chinese made ocean voyages to Africa, bringing back an enormous number of specimens of flora and fauna, including giraffes, which resembled a beneficent mythical creature, the *qilin.* During this period, the paddle wheel was adapted for use on military craft by General Li Gao (733–792), who constructed "fighting ships with two paddle-wheels on each side so that they ran as fast as swift horses."[15] Such ships were used on Dongting Lake, which converges with the Yangtze in Jiangsu province, but after a battle in 1135 when the rebel leader Yang Yao (?–1135) was defeated by General Yue Fei (1103–1142), who clogged the wheels of the battleship by throwing straw mats into the water, paddle boats were rarely seen again. By the fourteenth century, China had one of the most advanced navies in the world with a fleet of some 4,000 ships, but by the mid-fifteenth century, growing isolationism ended further exploration and advances in shipbuilding for over three hundred years.

Along the Yangtze, relatively simple ships were effective for transportation. The sampan and the wupan were the two basic boats in the gorges, the sampan floor being made of three long boards and the wupan five, large enough to carry a water buffalo broadside. Equipped with a tall sail stiffened by bamboo battens and a sculling oar, called a *yuloh*, the boat was usually manned by a crew of four, but often held at least a half-dozen other people as well. The sampan and wupan are examples of flat-bottomed riverine craft, believed to be little changed since ancient times. Commonly known as *kuaize*, they are still seen throughout the Three Gorges, though they are now rarely used for long-distance travel. G. R. G Worcester (1890–1969), the river inspector of the Chinese Maritime Customs on the Yangtze for thirty years, described over 250 junks used on every segment and tributary of the Yangtze in his comprehensive work on the subject. The junk still seen most frequently in the Three Gorges, though now more and more rarely, is called the *mayangzi.* This originated on the Yuan River in Hunan province and proved to be the most suitable craft for upriver Yangtze navigation. Worcester explains its history:

> Its fame spread to the middle Yangtze, when it is not known, on which sections of the river it became the recognized cargo-carrier. Later it found its way into Szechwan, in which province the first of its type to be built was constructed by a Hunan merchant who took up business in Chungking. It soon demonstrated its superiority over the other craft as the basic type best suited to the navigation of the rapids above Ichang, though retaining the old generic name of *ma-yang-tzu*, which explains why the old Hunanese name is still in general use in Szechwan for the most representative craft of the gorges, which is a modified and adapted form of the old prototype. This also accounts for the popular saying that none but a Hunan carpenter can build a true *ma-yang-tzu.*[16]

He adds, "All the *ma-yang-tzu* types are heavy, cumbersome, deep-draught cargo-carriers, strongly built to negotiate the long-distance voyages required of them. The sizes vary greatly from a length of 110 feet down to only 38 feet, but invariably the main features are faithfully adhered to."[17] In 1939, a new, lighter style of junk was introduced under the auspices of the Chongqing government. This was the first new design to appear, or at least to last, on the upper Yangtze in hundreds of years. Long and narrow, with a hull of cypress, it was built in twenty- forty-, and sixty-ton sizes. The motorized version is still a common sight along the river.

For over a thousand years, junks offered a lively variety of services to other boats and towns in remote areas all along the Yangtze. Actors and musicians performed operas, acrobatics, and magic acts on large theater junks that moored for a few days in the county seats and then sailed on. Floating hotels provided accommodations for people who arrived at night after the city gates were closed and no one was permitted to enter. Junks also served as tea houses and restaurants and grocery stores. From crowded sampans, farmers sold vegetables and pigs to passing boats. There were also the sing-song sampans, a tradition that continued well into the 1930s. Cornell Plant described the harbor at Fengjie in 1921, a vision hard to imagine for anyone who has visited the city recently: "Yet some romance still clings to it, and folks have even dubbed it the 'Venice of the Yangtze,' presumably on account of the many small gondola-like boats which have ply on the adjacent side stream, or perhaps because of the singing-girl sampans which ply up and down the tiers of junks at night serenading the hardy boatmen, who spend their hard-earned cash in the free and easy way peculiar to the sailing-ship sailors of by-gone days."[18]

In general, each boat, large or small, carried a large number of people. In contrast to a common Western tradition, there was no sense of bad luck attached to the presence of women on board. By many accounts, the wife of the owner was often a formidable presence. Although it was rare, as early as the nineteenth century, women, usually the widows of junk owners, ran commercial junks on their own. Almost all junks and smaller craft were worked by the owner himself, known as the *laoban* or *lao da* (literally, the boss, or old big). In addition to him and his family, a large wupan usually had an on-board crew of at least four men to work the sails and row, two pilots for navigation, fifteen trackers, including a drummer who beat out signals to the trackers on shore, and in the water, two swimmers specializing in disentangling underwater knots, and a cook. By the end of the nineteenth century, a large cargo junk, such as a 120-foot *mayangzi*, could carry as much as sixty tons upstream and eighty or ninety tons downstream, on a trip that averaged, if all went well, from twenty-five to sixty days between Chongqing and Yichang. Such boats usually employed about a hundred men, among them sixty or seventy trackers, and often hired additional trackers and pilots in difficult spots.

Life on a damp and crowded junk was uncomfortable and dangerous, but as long as the boat stayed afloat, the owner and his family were sure of a livelihood and had more freedom and stability than peasants who tilled the land, who not only had to depend on their crops and the weather, but also were often subject to the exploitation of landlords and tax collectors. Junks were taxed on the basis of draught or capacity, which made it possible to design boats to minimize the measurements that might cost the owner more. Of everyone on board, the cook was by far the most important person after the *laoban*, for he was not only responsible for feeding everyone, very important in Chinese culture, but for supplying the ship, mediating arguments, making sacrifices to the gods and spirits at the start of each journey and along the way, and preparing the requisite pork banquet at the end of a successful trip. The helmsman, or pilot, who had to navigate the craft through the rapids, ran a close second in terms of importance. In the eleventh century, Lu You made notes in his journal about the beginning of his trip through the Three Gorges, entries that could have been made almost any time up to the mid-nineteenth century.

> 17th day: After sundown we shifted our baggage to a boat belonging to Ch'ao Ch'ing of Chia-chou—the kind used for travelling through the gorges. Most of the people living along the embankment are natives of Shu or are married to people from Shu.
>
> 20th day: We have taken down the mast and prepared the fixtures for the oars. Going up the gorges, we will use only oars and "hundred-chang" towing lines and will not spread the sail. The "hundred-chang" are made of huge pieces of bamboo split into four strands and are as big around as a man's arm. The boat I'm riding on, of the 1,600-bushel class, uses six oars and two hundred-chang lines wound on winches. . . .
>
> 28th day: We tied up at Fang-ch'eng. There is a man named Wang Po-i from Chia-chou who earlier, in response to our request, agreed to act as *chao-t'ou* or chief helmsman of the boat; he was given special pay, and whenever an offering of meat was made to the gods, he received twice as much as the other men.[19]

In China, not only does every rock and inlet have a story associated with it, but a god as well. It is a disorderly pantheon, with gods and spirits from Buddhist, Taoist, historical, and animistic origins appearing separately and individually, often known in only one region or community, or responsible for just a fragment of a river or a particular mountain peak. On all rivers, boatmen pray to the river and mountain gods for assistance and for a safe journey, but exactly who or what these gods are is not always evident. While local descriptions make clear that some gods have a distinct human or animal form, others are portrayed more as a spirit or force. Gods and goddesses who were once human and fairies such as the Goddess Peak or Guan Yu, whose spirit protects sailors in distress, are generally helpful to the living, but the animistic spirits and gods of the river itself tend to be more mercurial and violent, and need to be constantly placated with sacrifices and offerings. The most dramatic example of this was He Bo, the ancient river god who demanded that every year two young women and two young men be sacrificed to him in the roaring waters of the upper Yangtze, near the border of what is now Yunnan. This practice was abolished in the fifth century BC.

The most popular god along the upper Yangtze was Yang Tai, a pirate on Dongting Lake who lived from 1127 to 1162 and came to be known as Zhen Jiang Wang, the River Protecting God. The young engineer in John Hersey's novel *A Single Pebble* describes his origins:

> The river-guarding god, Chien-chiang Wang, was not, it seemed, a being imagined by river folk in mystery, fear, and yearning, but was in fact the apotheosis of a human rascal, the deification of a twelfth-century pirate of the Great River whose only claim to divinity, Su-ling had told me, was unending success; he could not fail in his mischievous undertakings—until, at last, meeting defeat in very old age, he plunged into the waters of Tungting Lake and drowned himself. So perhaps these roguish priests were suitable after all; and perhaps, it occurred to me, this wanton traffic on the wild rapids and in the profound gorges of the great river was a kind of mischief, and perhaps it was fitting to ask success in it from a master knave.[20]

The sacrifice to the river god required cutting the throat of a cock and then sprinkling blood on the sides of the boat, exploding firecrackers to frighten off evil spirits, and burning incense. Once the cook or pilot accomplished this, the junk was ready to depart and everyone hoped for the best. Control of the river was both desired and feared, much wanted by boatmen and farmers along the shore but not at the cost of upsetting the gods by changing the river or doing anything else that might anger them. They hoped to calm the river dragons, not to harm or still them.

From ancient times to the present, the struggle between living in harmony with nature or controlling it has been a continuing philosophical controversy. For centuries, China's emperors attempted to harness nature to achieve their goals, creating some of the world's biggest irrigation projects and man-made waterways, but the limits of technology meant that relatively little permanent destruction to the environment resulted from the process of ordering it to suit man better. In the third century BC, Mencius (ca. 372–289), one of the great Confucian philosophers, recommended a practice of conservation and balance. "If the seasons of husbandry be not interfered with, the grains will be more than can be eaten. If close-meshed nets are prohibited in the pools and lakes, the fishes and turtles will be more than can be consumed. If axes and hatchets are used in the mountain forests only at suitable times, there will be more wood than people know what to do with."[21] The balance that had persisted for lack of the means to undo it completely was set askew by the introduction of modern technology, political upheaval, and international commerce. The Yangtze has always had disastrous floods, but in the late twentieth century these occurred even more frequently. Thousands of years after the first river control projects in China, there are still people in the towns in the Three Gorges who remember the gods of the river and mountains. They wonder if the frequent destructive floods of the late twentieth century were truly only a natural consequence of dike building and deforestation or perhaps a more complex and active revenge by nature and whatever spirits inhabit it to punish man for his excessive interference.

CHAPTER 4

OUTSIDE PRESSURES

FROM THE EARLIEST WRITTEN RECORDS OF TRAVEL along the Yangtze, which date from the poems of Tang dynasty scholar-officials, the tension between the local inhabitants of the upper Yangtze and the people who came from the outside is evident. As far back as the third century BC, China's rulers sent subjects they wanted at a distance into exile in Sichuan. Few people made the long and hazardous trip to the deep interior voluntarily and those who did, except for missionaries, went to the mountainous villages and towns of the gorges to retreat from the world rather than to bring change. Only in the late nineteenth century did outsiders begin to penetrate the upper Yangtze in significant numbers and for more diverse reasons. At this time, as the Qing dynasty weakened, the number and intensity of domestic and foreign conflicts soared, resulting in the presence of both Chinese and foreign military along the river. Lured by the expansion of international trade, European and American merchants came to take advantage of new business opportunities. Travelers also began to arrive for the sake of adventure, a concept scarcely known in China at this time and one still largely incomprehensible to local people.

Regardless of the century, visitors to the upper Yangtze River and the inland villages (or at least those who left a written record) have responded with surprising similarity despite the thousands of years that have separated their trips. New arrivals often express a mixture of awe at the power and treachery of the landscape and consternation at having ended up in a place so unfamiliar and backward, populated by a people with peculiar customs and with whom they could scarcely communicate. In the Tang dynasty, Li Bai (701–762), one of the most important of Chinese poets, described the Sichuan Road in a poem of the same name. In this excerpt, he wrote of the plank and trestle pathway that led over three hundred miles of cliffs from the Yangtze to Chengdu:

> This is a fearful way. You cannot cross these cliffs.
> The only living things are birds crying in ancient trees,
> Male wooing female up and down the woods,
> And the cuckoo, weary of empty hills,
> Singing to the moon.
> It is easier to climb to heaven
> Than take the Sichuan Road.
> The mere telling of its perils blanches youthful cheeks.
> Peak follows peak, each but a hand's breadth from the sky;
> Dead pine trees hang head down into the chasms,
> Torrents and waterfalls outroar each other,
> Pounding the cliffs and boiling over rocks,
> Booming like thunder through a thousand caverns.
> What takes you, traveller, this long, weary way
> So filled with danger?
> Sword Pass is steep and narrow,
> One man could hold this pass against ten thousand;
> And sometimes its defenders
> Are not mortal men but wolves and jackals.
> By day we dread the savage tiger, by night the serpent,
> Sharp fanged sucker of blood
> Who chops men down like stalks of hemp.[1]

Given the difficulty and unpleasantness of the trip, it is not surprising that prior to the nineteenth century the only Westerners known to have ventured into this area were a few Jesuit priests. When the Yongzheng emperor outlawed the Catholic church in China in 1724, a move generated by his fear of its potential political power, almost all foreign priests were required to leave the country. At this time, there were an estimated 300,000 Catholics in China, many in Sichuan. The number declined sharply by the end of the eighteenth century, but pockets of Catholicism survived in Sichuan, which had had one of the largest groups of converts. Christianity surged again in the towns near the Three Gorges with the reintroduction of missionaries in the mid-nineteenth century.

The roots of the massive political and social change in China that led to the end of the imperial system and eventually the Communist victory reach back to the decline of Qing power in the nineteenth century and its increasing but unwilling interaction with other nations. Between 1840 and 1940, internal and international conflicts that began in Peking had consequences that reverberated around the country. Along the upper Yangtze, the impact of this period was both immense and nonexistent. In the 1840s it still could take six months or more to reach Chongqing or any other place beyond the gorges from almost any important city in China, and few people had reason to make the trip. Peasants hacked tiny fields out of the mountainside with wooden hoes and watered their sorghum and potatoes with buckets of water carried from the river or village wells over mountain paths. The majority of rural people had never met anyone from farther away than the next county, and Beijing was an almost imaginary place where the emperor lived. By 1940, Chongqing had become the wartime capital of China, with a population that had grown in three years from a few hundred thousand to two million. The steamship had revolutionized transportation, and foreigners from all over the world were floating up and down the river, taking for granted that this right was theirs. An endless continuum of wars had disrupted millions of lives. Meanwhile, peasants still terraced their plots and held temple fairs and prayed to the river god and somehow survived. Despite the importance of this era, it is almost as hard to find traces of it in this region as it is impossible to escape reminders of fourth-century battles or the political struggles of the Tang dynasty. Ancient history may be alive in river towns like Fengdu or Zigui, the tales a part of everyday life, but modern history is elusive.

Three things—the spread of opium, China's opening to the West, and the introduction of the steamship a half-century later—had an enormous impact on life along the Yangtze River, but they are barely remembered here. Ask anyone in Fengjie about the battles of the Three Kingdoms and he or she will talk for half an hour. Ask about the warlords of the 1930s, who struggled for control of the same territory, and where they fought or how they supported their armies, and most people younger than about eighty will look blank, or say something like, "*Bu qingchu. Dagai hen luan.*" (I'm not sure. It was probably chaotic.) While the distant past gives a sense of security and valor, more recent times often point to weakness or chaos, or worse, bring up uncomfortable questions about who is responsible for what. Yet, parts of this past keep creeping into the present, affecting and shaping what the region has become.

It is possible to give here only a brief introduction to the forces that began to push one of the most remote and independent areas of China into wholesale engagement with the rest of the world. One catalyst was the Opium War (1839–1842) with England, which everyone in China—in Beijing, Guangzhou, and every town along the Yangtze River—is taught was the beginning of the oppression of China by foreign powers. It is an historical event of immense importance, the beginning of a time of national shame. People's grasp of the details is fuzzy—surprising only because of the clarity with which earlier times are recalled. Thinking back on American and European history classes in which no one ever reached World War I before the end of the year, let alone the present, I asked a Chinese friend why there is such a vague understanding of the last few hundred years of history. He said that scholars and other experts with a special interest know history well, but for ordinary citizens it is not so important. "If the history is not exciting and glorious, they find it boring. What does it have to do with their lives anyway?" he asked, adding the plaint of someone reaching middle age, "I like to learn about the past, but young people only want to make money."

Since the beginning of its contact with Western merchants, the Qing government was hesitant to promote international trade, allowing foreign traders to live only in certain ports (initially only Canton) and to work only through designated Chinese middlemen. The Qing refusal to permit open trade and free access was a long-term source of friction between China and England in particular. Beginning in the mid-eighteenth century, the British controlled a flourishing three-way trade route from England to China and India. Britain exchanged its own products and Indian cotton for Chinese tea and silk, which were then sold in England and Europe. Within a few decades, the British demand for tea and other goods exceeded the limited Chinese desire for anything English, resulting in a growing British silver deficit as merchants were forced to pay for their purchases in cash. To redress this trade imbalance, in about 1800 the British began to export enormous quantities of Indian opium to China to create a market for something the British could supply.

Although opium had been known and used in China since the seventeenth century, smoking it was prohibited in 1729 and its cultivation and importation officially but ineffectually banned in 1800. From the early nineteenth century on, the massive increase in supply brought in by the British, coupled with other social disruptions, caused an overwhelming increase in the number of addicts. The Qing attempted to curb this, issuing other edicts prohibiting the use of opium and officially closing the port of Canton to opium imports, with little success. Between 1829 and 1839, the British exported an average of over 1,840 tons of opium per year from British India to China, an amount comparable to the average annual opium yield in Burma (the world's second largest opium producer) in the late 1990s. In 1839, the Chinese government attempted once again to take a decisive stand against the drug and burned over 20,000 cases of opium from British ships docked in Canton. This act, along with continuing tensions with England about free trade, led to the Opium War, which China lost. The resulting Treaty of Nanjing in 1842 required

China to open five coastal cities (Guangzhou [Canton], Shanghai, Xiamen, Fuzhou, and Ningbo) as treaty ports, where foreigners were permitted to trade and live subject only to their own laws. China also agreed to pay Britain an indemnity of 21 million dollars for the loss of the destroyed opium and other damages and to cede Hong Kong Island to Britain in perpetuity.

Later treaties, including the 1858 Treaty of Tianjin negotiated by Lord Elgin and the 1876 Treaty of Chefoo, established the most-favored-nation system, under which all signatories to a treaty with China received the same rights granted to any other treaty partner, and set low tariffs on foreign imports. Treaties permitted foreign military as well as commercial transportation along rivers, allowed the return of missionaries, and gave foreigners the right to travel inland with a government guarantee of safe passage. They also increased the number of treaty ports, which eventually grew from the original five in 1842 to nearly eighty over the next five decades, including Zhenjiang, Nanjing, Hankou, Shashi, Yichang, and Chongqing along the Yangtze.

The 1858 Treaty of Tianjin imposed a tax on imported opium for the first time and began a forty-eight-year period of toleration and de facto legalization of importation and domestic cultivation of opium. After 1860 the amount of opium imported from British India continued to grow, reaching a peak of 6,500 tons in 1880. Imports then decreased because of rising domestic production and China's mounting financial difficulties. Chinese opium was cheaper than Indian opium and the quality quickly improved. Easy to transport and with a wholesale price two to three times that of wheat, it became a lucrative crop for peasants to grow. Opium permeated Chinese society on a scale that is hard to imagine. For more than a century, from the mid-1840s until the 1950s, opium production and addiction were major political and social issues throughout China, affecting the health and productivity of millions of people.

In the Yangtze River cities, if one looks hard enough, one can glimpse how faraway events in Peking and London opened the Three Gorges to the outside. Thomas Blakiston (1832–1891), a naturalist and explorer, had served in the Crimean War and was sent to command an artillery detachment in Canton. From 1860 to 1862 he explored the middle and upper Yangtze entirely by junk. In 1862, Blakiston described the expanding opium crop near Fengjie, one of the results of the Tianjin Treaty of 1858:

> It was in this neighbourhood that we first observed the poppy cultivated, and hence onwards it was very common; and, from the amount which we saw along the banks of the river, it would appear that the quantity of opium raised in Sz'chuan must be very large. In the same patch one sees pink, lilac, and

> white flowers, and the appearance of the beds of poppies on the terraces of the hill-sides among the other crops is very beautiful. When the flower dies off, the seed-pod, or head, is scored with several cuts vertically, from which oozes a substance of the appearance of freshly warmed glue; this is collected by the farmers and their families, who scrape it off with a knife and deposit it in a little pot which each person carries for the purpose. . . . Yet all this cultivation—for it is said to be also grown extensively in the south-western provinces—and consumption of opium, are in violation of the law, and furnish only another instance of the universal state of decay of the government of this wonderful country, where . . . pipes, lamps, and all the apparatus for smoking opium, are sold publicly in every town, and the mandarins themselves are the first to violate the law and give this bad example to the people, even in the courts of justice.[2]

By the end of the nineteenth century, opium had become one of the main sources of revenue of the provincial governments of both Sichuan and Yunnan provinces. Opium was Sichuan's major export crop (followed by silk, salt, sugar, and medicine), with two-thirds of it distributed elsewhere in China. In the 1904 "Report by Consul-General Hosie on the Province of Ssu'chuan," the former British consul in Chongqing stated that "in cities 50 per cent of the males and 20 per cent of the females smoke opium, that in the country the percentage is not less than 15 and 5 per cent, respectively. . . . While the British connection with the trade was still indicated by the term 'foreign smoke,' practically no foreign opium was used in this province; on the contrary, vast quantities were exported to other parts of this empire."[3]

By the early twentieth century, there were an estimated twenty million opium addicts in China. In 1906, the Chinese government launched a campaign to suppress opium. This was one of the most successful, if short-lived, of the late Qing dynasty reforms, but after the fall of the Qing in 1911 following a rebellion in Wuhan, opium growers resumed large-scale production. By 1916 opium was again one of the economic mainstays of both Sichuan and Yunnan. Throughout the 1930s, opium sales funded the warlord armies that fought against one another throughout the provinces, and softened life for soldiers pressed into service in wretched conditions by whatever army happened to be in the neighborhood.

In the early 1950s, the new Communist regime launched a harsh and effective campaign against opium use, breaking millions of people of their addiction and imprisoning or executing close to 80,000 large-scale opium producers and drug dealers. After this, narcotics seemed to have been almost entirely eliminated for decades, but in the 1990s, with open borders and a new ease of travel, almost-forgotten problems made their way back. In the karaoke clubs of Wushan, dance bars of Wanxian, or discotheques in old air-raid shelters in Chongqing, young people listen to Taiwanese and Japanese music and roll their heads and twitch their feet. Some are on a form of methamphetamine called *yaotouwan* (the rock-your-head drug), others just pretend to be. "It's cool to look like you are taking *yaotouwan*," explains one 24-year old, "but it's much cheaper if you just shake your head." Other young men tell stories of friends, the children of workers or educated parents, who took up with the wrong people, began shooting up with heroin, and came to a bad end. They seem mystified by this, unable to grasp whatever leap it takes to stick a needle in your arm and risk an encounter with China's draconian narcotics laws. A junior employee of one of the passenger boats says it's better than it was once. "Back in '94 or '95, when you went to a party, people would pass out pills, and everybody wanted to try them. Everyone thought of them as medicine that made you feel good. Nobody connected them with drugs." Western visitors report teenagers bragging to them on the streets of Wanxian about how easy it is to get a bag of heroin. Travelers who have ventured farther afield say that deep in the hills behind the gorges violet fields of poppy bloom again.

In the nineteenth century, the treaty rights granted to foreigners, including inland travel, military and commercial river transportation, and the return of Christian missionaries, resulted in a sudden influx of Westerners along the Yangtze. Dozens of detailed and colorful accounts by Western merchants, adventurers, and missionaries who sailed the Yangtze during the early post-treaty era provide a wealth of information about local conditions and the difficulties of dealing with the people and the geography. Unfortunately, there are few such documents recording the local reaction to these strangers in their midst, which was not always enthusiastic. The China Inland Mission, an American Protestant missionary group established in 1865, began to make inroads in Sichuan in 1877. According to R. J. Davidson, the author of a 1905 book on western China, the mission's goals were straightforward—"one aim and purpose was the evangelization of the whole Empire in the shortest time possible, rather than obtaining the largest number of converts in a limited field."[4]

Once the China Inland Mission gained a foothold in Chongqing, other groups quickly followed into more rural areas. These included the American Presbyterian Mission, the Methodist Episcopal Mission, the London Missionary Society, the American Baptist Missionary Union, the Friends' Mission, the Church Missionary Society, the Canadian Methodists, and the Scotch National, British, American, and Foreign Bible Societies. The impact of the missionaries, who often went into remote areas, was significant on a regional level and frequently resented by officials who feared that the churches would undermine their authority. Peasants and townspeople, even those who themselves became churchgoers, often did not entirely understand or trust the motives of the foreign ministers and priests. Davidson, one of the 200 missionaries in Sichuan at the end of the century, describes the Sichuanese as "the most gentle and amiable in character, and most refined in manner. They are also more cleanly and orderly in dress and habits than the Chinese in general. They meet you with civility and a show of respect, and are not in the least shy, they answer frankly without hesitation."[5] Despite his admiration, such good feelings were not always reciprocated, and there were frequent conflicts, particularly over the use of farmland to build churches. In 1863 and in 1886 there were antimissionary riots in Chongqing, known as the first and second "church incidents," the first directed against Catholics, the second against Protestants, with resulting tensions along the river running high enough that Western travelers were warned not to enter the cities.

The Catholic church also returned to China, with French missionaries fanning out into the countryside in the 1840s, joined by Americans and other nationalities twenty years later. In the Three Gorges region, the center of church activity was the town of Damiao, on the opposite shore from Wushan, about twenty miles inland. According to a lay sister at the Catholic church in Wushan, it once had a congregation of more than 30,000 people. Western travelers of the time described the steeples of the fine cathedrals and churches you could see from the river in almost every city along the Yangtze, among them the grand cathedral in Wushan, built by French missionaries in 1898.

A number of the old Christian churches remain along the river. Those that are still intact give the visitor a sense of a bygone time almost entirely lacking in towns where most buildings are either made of poured concrete or are supposedly at least a thousand years old. In Fengjie, a simple and elegant Protestant church, said to have been built by Northern Baptists from New England at the turn of the twentieth century, still stands just inside the city wall, largely unscathed despite its decade as a sulphuric acid factory during the Cultural Revolution. Like all Protestant churches in China, it is part of the Three-Self Patriotic Movement (self-supporting, self-governing, and self-propagating), under which all Protestant denominations were combined after 1949, leaving the individual churches to evolve out of the half-forgotten vestiges of whatever group was practicing here in the past.

In neighboring Wushan, the original nineteenth-century Catholic church building did not fare so well. High on the bluff above the river, just

Figure 4.1 Catholic church in Tianchi, Zhongxian county

inside the city wall, the cathedral was located on the main street overlooking the river, a reminder of the power that the Christian churches once had to obtain prime real estate. The location also points to the importance of good *feng shui*, a necessity insisted upon by Chinese members. In Wushan, however, the correct positioning of the building in relation to the water and mountains was not enough to protect it from the forces of history, and the church has a record of harsh conflict with the local government since the Communists came to power.

Like other Chinese Catholic churches recognized by the government, the Wushan congregation is a member of the Chinese Catholic Patriotic Association, which by government mandate does not acknowledge the pope as the head of the church, though individual members often quietly disagree. The site of the original cathedral was restored to the church only in the mid-1980s after a relentless campaign by a woman known as Yang Xiansheng, or Teacher Yang, a lay leader of the church. She is a forceful and stubborn woman in her nineties, who still keeps track of an extensive program of good works and daily religious services. Because of the shortage of priests, masses in the former factory are held only once a month, but daily services led by lay members are packed at five in the morning and again twelve hours later. Throughout the 1990s, church members negotiated with the city government about the location of the new post-flood church and fought for greater compensation. Plans have since been completed for a new complex that will include not only the church but also a daycare center and guest house.

Western nineteenth-century writers described the local temples and pagodas that flourished along the river at this time. In his book *The River of Golden Sand: The Narrative of a Journey Through China and Eastern Tibet to Burmah*, William Gill (1843–1881), who traveled through the Three Gorges by junk in 1868, portrayed Shibao Block as it was then, before the destruction of the pagoda's Buddhist figures during the Cultural Revolution a hundred years later.

> We walked in the afternoon to Shih-Pao-Chai (Stone Jewel Fort), where there is a very curious rock, the top of which is about 150 feet above the ground on which it stands, and 300 feet above the river. All the faces of it are perpendicular, and on one side a pagoda has been built against the rock. We visited this, and went to the top, where we found a number of wooden figures carved and brilliantly painted and gilded. These figures represented the horrors of hell, and the punishments reserved for the wicked; the judges were life-size figures of men, with long black beards and mustachios, holding books in their hands. The executioners were horrible devils, with hideous

faces, who were supposed to be thoroughly enjoying their odious tasks of torturing the victims. These tortures were of many kinds, and the most abominable that it is possible to imagine: Some of the sinners were being sawn in two; some were being ground in a mill; and others were arranged in layers between big flat stones, with their heads only projecting all round, while a demon stood above spearing their faces. There was no representation of Heaven. The eighteen saints with twenty-four attendants were in another room; and at the entrance, the god of literature, a repulsive creature, was standing on the point of one toe like a ballet dancer.[6]

Close to thirty years later, Mrs. John Bishop (1831–1904), an energetic and eccentric British woman explorer, visited Wushan and Fengjie. Known also by her maiden name, Isabella Bird, she was married late in life to a patient suitor not put off by her frequent trips to China and central Asia, which often lasted several years. In *The Yangtze Valley and Beyond: An Account of Journeys in China, Chiefly in the Province of Sze Chuan and Among the Man-tze of the Somo Territory*, Mrs. Bishop described Wushan and Fengjie as dreary communities with isolated spots of interest or beauty, not so different in some ways from the present. She wrote of Wushan:

[It] is grey and picturesque, its walls following the contour of the hills on which it is built, enclosing fields, orchards, and beautiful trees. A fine temple to the God of Literature in a grove of evergreens on a steep mountain cone 1500 feet in height, and a lofty pagoda on the same peak are striking objects, but the town, though fairly clean, has no look of prosperity, and so far was disappointing.[7]

In Fengjie, then known as Kuei Fu, she noted the growing problems between local residents and foreigners as a result of treaty-imposed changes in shipping and taxation.

Kuei Fu . . . is a decaying city, bolstered up into an appearance of grandeur by its position and its stately wall and gate towers. There all goods going up or down the Yangtze paid *likin*, a transit tax of about 5 per cent. on their value. As (according to Mr. Little) over 10,000 junks go up and down in the year, and each one is delayed for examination three or four days, a large extramural population made a living by supplying their needs. Some years ago the Kuei Fu Likin Office was the most valuable in China next to that of Canton, and the likin duties were the great source of SZE CHUAN revenue. The grand houses, with fine pleasure grounds, of which many can be seen from a height above the wall, testify to the fortunes made by officials in the days

> when they had the right to levy 5 per cent. on a trade worth possibly £2,000,000 pounds sterling.
>
> But we have "changed all that" by securing the opening of the treaty port of Chungking with the transit pass and chartered junk system, to which all foreign imports can be carried on payment of duty to the Imperial Maritime Customs at Shanghai. Thus these riches go to Peking, and the "Four Streams Province" is the sufferer, and Kuei Fu really can only exact legal dues from junks carrying local merchandise and from salt junks. The reader will at once perceive the reason for the strong provincial hostility which is aroused by the opening of new treaty ports, for each one, to a greater or lesser extent enriches the Imperial Government at the expense of the provinces and deprives a great number of officials of their "legitimate" perquisites or "squeezes" in favour, as the people think, of highly salaried foreign customs employés.[8]

The tensions Mrs. Bishop describes were to become more severe over the next two decades, with the introduction of reliable steamship service. Fengjie was hard hit as it went from being the main source of tax revenue in Sichuan province to an impoverished city. In 1897, the collection of taxes on all foreign shipments on the upper Yangtze was transferred from local agents in Fengjie to the Imperial Maritime Customs in Shanghai, with devastating effects on the city, removing huge amounts of money, legal and otherwise. At this time, there were an estimated 10,000 junks on the upper Yangtze, employing between 250,000 and 400,000 men, work that was lost as steamers took over passenger and freight transport in the following decades. In addition, dozens of trackers, men harnessed by tow-ropes to the boats, might be needed to move a single large junk upriver, and this number could reach the hundreds when it came to maneuvering larger ships through the rapids. The trackers' work also gradually disappeared as modern vessels able to negotiate the rapids without assistance became more common along the upper Yangtze. The work of the trackers was difficult and dangerous, but for landless peasants, providing the strength to drag ships through the upper river was one of few ways to earn a living. Though junk owners entrusted the fate of their craft and their crew to the skill of the trackers and their knowledge of the rocks and currents, people on land saw tracking as one of the lowest forms of labor, a menial job of last resort for men who could do nothing else. Many Western authors in the late nineteenth and early twentieth century looked at the life of the tracker with a combination of pity and awe, and sometimes with a sense of romance entirely foreign to the Chinese view of the work. An excerpt from John Hersey's novel *A Single Pebble* describes Old Pebble, the head tracker.

He was, she said, the strongest man she had ever seen. She said there had been times, when unexpected freshets had come, when he had hauled from sunup to sundown without stopping for so much as a cup of tea.

I asked what he worked so hard for.

"He works for work," she said. "He loves the work for its own sake. He loves everything he does for its own sake—everything he does on the river, that is. When he is in a city he is a poor man, he is like a crippled beggar."[9]

Mrs. Bishop focused on the specifics of the danger:

At some points where the rapids are bad and the shores are big broken rocks, only fitted for goats to climb, and the junks hang or slip back, and the men give way, and several big junks, each with from 200 to 300 trackers, are all making the slowest possible progress, gongs and drums are beaten frantically; bells are rung; firearms are let off; the hundreds of trackers on all fours are yelling and bellowing; the overseers are vociferating like madmen, and rush wildly along the gasping and struggling lines of naked men, dancing, howling, leaping, and thrashing them with split bamboos, not much to their hurt. A tow-rope breaks, and the junk they are tugging at gyrates at immense speed to the foot of the rapid, the labour of hours being wasted in two or three minutes, if there is not a worse result. . . .

No work is more exposed to risks of limb and life. Many fall over the cliffs and are drowned, others break their limbs and are left on shore to take their chance—and a poor one it is—without splints or treatment; severe strains and hernia are common, produced by tremendous efforts in dragging, and it is no uncommon thing when a man falls that his thin naked body is dragged bumping over the rocks before he extricates himself.[10]

With the blasting of the worst rapids along the Yangtze in the 1950s and 1960s, tracking as an occupation came to an end. Still, along its tributaries, on the Shennong River near the border of Hubei and in the narrow waters of the Little Little Three Gorges (the even smaller gorges on the Madu River, a tributary of the Daning River near Wushan) men pole and pull small boats through the fast-moving shallows. These open boats serve as ferries that take inland villagers to visit relatives near the Yangtze or bring tourists to interior rivulets for activities such as rafting on rubber tires on the gentler rapids. While the men who work here are rarely in danger, and they use their poles as they walk in the water, rather than maneuvering the boats from a crouching position high on a cliff, it is still hard work. An excerpt from an eighth-century poem by Wang Jian (768–833) still expresses how the men who pull boats feel about their work, something that has not changed much in the past 1,200 years.

> Oh, our life is bitter by the cargo station,
> We are forced by officials to pull the heavy boats.
> Day and night, seldom do we find happiness.
> At night we sleep on the water, by day, we walk on the sand like seagulls.
> Against the wind, facing upstream, our load is as heavy as ten thousand bushels.
> The next station is far away; the last one invisible among the waves.
> Sometimes we must work at night
> Despite heavy snow and torrential rain.
> In the cold night, our thin clothing is soaked
> And our bare feet chapped by the frigid damp.
> Working till dawn, whom can we tell of our misery?
> All we can do is sing as we pull.
> Our shabby houses are worth little
> But we will not abandon our homes.
> We wish this river would turn into plots of farmland
> And then we would not curse our difficult lot.[11]

In the Little Little Three Gorges, the mountains are steeper and higher than those closer to the Yangtze. Looking upward, you see no sign of human habitation, but somewhere behind the hills are houses and family plots. At first it is impossible to imagine how anyone might reach one of these invisible places, but as you focus your eyes on the greenery, tiny streaks emerge that turn into mountain paths. One of these leads from the roof of the only hotel in the area, built in 1997 in the hope of attracting more visitors to the area. A leap from the hotel roof to a foot-size ledge puts you in a position to begin the climb on a twelve-inch-wide dirt path that careens upward at a terrifying angle. You begin your walk clinging to the edge of a cliff hundreds of feet above the river and end it somewhere, thousands of feet higher up, a destination that the first ten feet of the journey convinced me I would not be reaching.

Some men here own boats with brothers or cousins, and live on the riverbank for part of the year, leaving their families in the hills to farm. The men have had hard lives at barely subsistence level and they are still poor, but this is changing somewhat for the dozen or so who have developed a kind of tiny monopoly pulling the occasional tourist boats. The boatmen ham it up for visitors, get photographed by foreign journalists, and trade cigarettes with young entrepreneurs from Shenzhen and Hong Kong who have somehow taken a detour from the usual sights. The men still sing the traditional songs of the

Figure 4.2 Men pulling boats in the Little Little Three Gorges of the Madu River

river, pushing and pulling the boats through the clear water with a thumping rhythm. When the dam is completed, the water level, now three to four feet deep on average in midsummer, will be forty feet deep. The mountains will be just as high and no easier to grow anything on. As the local people already live in the mountains because there is so little flat shore, the higher water level here will require few villagers to move, but it will not improve their lives or do much to alleviate the poverty. Local authorities are pinning their hopes, somewhat fancifully, on increased tourism once the narrow tributaries are flooded enough to allow the passage of larger and more powerful ships. The boatmen don't believe any of this. They want to show us their homes to prove how poor they are and that they should be paid more, but an official from Wushan tells them they are lucky to be doing this healthy work in the fresh air and that they get paid more than enough already. The men breathe in another couple long drags of tobacco and leap back into the river and push onward.

Archibald Little (1838–1908), an English pioneer in steamship design and navigation, did for the upper Yangtze what the local travel authorities hope larger motorized boats will now do for its inland tributaries. In the 1880s, it was cheaper to ship a ton of cargo from Hankou (now part of the city of Wuhan) to London than from Hankou to Chongqing. Little changed that by paving the way for the commercial passenger and freight service through the Three Gorges. Little, who had been in China since 1859 in one business scheme or another, made his first voyage through the Three Gorges in 1881 on a chartered junk. His trip convinced him that steamship travel through the Three Gorges was possible and he set out to make it happen. In 1884, Little introduced steamship service between Wuhan and Yichang, and in 1887 he founded the Upper Yangtze Steamship Navigation Company in London. Two years later, Little had a 175-foot ship, the *Kuling*, built in England and assembled in Shanghai. The *Kuling* sailed as far as Yichang in 1889, where it sat while Little spent two years negotiating with Chinese authorities for permission to sail farther upstream, which he never obtained. Though Chinese merchants and traders welcomed the advent of faster transportation, regional government officials feared a loss of financial control (which was often closely tied to their personal well-being) over this section of the river and the disruption of traditional ways. The government objected to the maiden voyage of the steamship along the upper Yangtze lest it be successful and thus followed by other such ships that would displace the junk. Concerns cited included the probable loss of work for thousands of junkmen and trackers, the potential loss of transit duties, damage to small craft along the river by larger, more powerful steamships, and exacerbation of already existing antiforeign feelings. Officials also noted that the new ship would upset the wild monkeys that lived on the cliff and would disturb the river spirits. All of this later came to pass to one extent or another.

In the end, Little gave up and sold the *Kuling* to the Chinese government, which resold it for a substantial profit to the Chinese-owned China Steam Navigation Company, where it was used for many years as a transport ship between Yichang and Hankou. Little's next attempt was the 50-foot *Leechuan*, a much smaller craft that did not arouse such anxiety among Chinese authorities. This boat left Yichang in mid-February 1898 and arrived in Chongqing three weeks later, having been pulled through most of the rapids by over a hundred trackers. Despite such problems, Little was certain that he could establish regular steamship service on the upper Yangtze and convinced Cornell Plant, formerly a ship's captain on the Tigris and Euphrates, to pilot one of the new steamships he planned for the Yangtze. Not long after, in June 1900, Plant sailed the *Pioneer*, a 180-foot-long paddle steamer, from Yichang to Chongqing in just seventy-three sailing hours.

Permitted by treaty to sail on the Yangtze after 1858, foreign warships, mainly British and American, had become commonplace between Shanghai and Hankou by the early 1870s. At the turn of the century, the foreign fleets began to penetrate the gorges as well in specially designed smaller gunboats. After Little's *Leechuan*, the next steam-powered ships to reach Chongqing were two British gunboats, the *Woodcock* and *Woodlark*, in May 1900. Their upstream trip took thirty-one days from Yichang and the assistance of 400 trackers. The Chinese made no further protests about steamship travel, and Plant's successful first voyage took place only a month later with much less trouble. Despite this, many Chinese and foreigners alike had doubts about whether this would ever be a practical means of transportation. Suspicions that it was a foolish idea were strengthened later that year when in December a German firm hoping to make commercial inroads in western China launched the *Suihsing*, a 200-foot passenger and cargo ship, huge for its time and place. Their plans ended quickly when it sank on its first upriver trip. Though local rescue crews saved hundreds of passengers, the accident put a damper on other such ventures for some years to come. The French had better luck sailing their first gunboat, the *Olry*, through the gorges in October 1901. The French navy went on to complete a detailed and at that time unrivaled charting of the upper Yangtze between 1901 and 1903.

Chongqing was opened as a treaty port in 1890 by an amendment to the 1876 Treaty of Chefoo, and the British soon established a consulate and trade mission there. Political troubles and problems in financing delayed the onset of regularly scheduled commercial steamship service along the upper Yangtze, which finally began in 1905 with the *Shuting*, a 115-foot ship with a specially designed cargo and passenger barge attached. Within a few years, bimonthly passenger and freight service through the gorges was running relatively smoothly. After Little's death in 1908, Cornell Plant raised money from Chinese sources and founded the Sichuan Steamship Navigation Company. He too remained in China until his death, becoming a legend along the river. He served as river inspector for the Chinese Imperial Maritime Service and designed the signal system still in existence along the Yangtze today: the dark arrows on signposts near little white houses above the riverbank that alert river captains to oncoming vessels. As in the past, downstream bound ships, which have less control because of the speed of the current, have the right of way. When Plant retired in 1911, the Chinese government built a cottage for him and his wife on the mountainside in Xiling Gorge, the most dangerous of the rapids, as an expression of gratitude for his services. The cottage is long gone and the rapids were blasted away in 1955, but a pillar commemorating him remains. Protected by local people in the village of Beishi during the Cultural Revolution, the memorial will move with them once the area is flooded by the Three Gorges Dam. Pu Lantian, as Plant was known in China, is still well known to the men who work the river.

Little, Plant, and the foreign and Chinese engineers and entrepreneurs who followed them permanently changed ordinary travel and international trade along the upper Yangtze. By the 1920s, over a dozen companies offered passage on the Yangtze, including the American-owned Robert Dollar Line, the most luxurious and largest of the foreign companies, and the Chinese-run Minsheng. As he was struggling to establish the steamboat, Archibald Little offered a convincing argument for its introduction. He wrote:

> . . . it must be borne in mind that the Yang-tse is not only the main, but the sole road of intercommunication between the east and west of this vast empire. Roads, properly so called, do not exist in China; narrow footpaths alone connect one town and village with another, and, except by the waterways, nothing can be transported from place to place but on men's backs. In the far north, it is true, cart-tracks exist, and clumsy two-wheeled springless carts are there in use, but in central and southern China, land travel is absolutely confined to paths, so narrow that two pedestrians have often a difficulty in passing each other. Traces of magnificent paved roads, of the ancient dynasties, still exist in nearly every province; but they have been destroyed by neglect, and have been disused for centuries past.[12]

The 1876 Treaty of Chefoo had permitted the establishment of foreign consulates in Chongqing but did not allow direct foreign trade, leaving this as a possibility for later, should the city ever be accessible by steamboat, which Chinese authorities at the time hoped and believed to be impossible. The Treaty stipulated that "The British Government will, further, be free to send officers to reside at Ch'ung K'ing to watch the conditions of British trade in Ssu-Ch'uen. British merchants will not be allowed to reside at Ch'ung K'ing, or to open establishments or warehouses there, so long as no steamers have access to the port. When steamers have succeeded in ascending the river so far, further arrangements can be taken into consideration."[13] With those further arrangements unstoppably in place by the 1890s, the upper Yangtze was slowly but irrevocably opened to storms from the outside. A Chinese proverb states, "When the Empire is peaceful, Sichuan is the first place to be in disorder; after peace is restored, Sichuan is the last place to be stabilized." By the turn of the century, the upper Yangtze, along with much of the rest of China, was in for a period of little rest.

CHAPTER 5

Change from Within

THE FALL OF THE QING DYNASTY IN 1911 BROUGHT an end to over two thousand years of imperial rule in China. Since the beginning of the dynasty in 1644, Chinese men had been required to wear their hair in a long braid, or queue, as a sign of allegiance to the Manchu rulers of the Qing empire. With their departure, the braid also had to go. Tan Shihua, the son of a revolutionary-minded government official, described his boyhood memories of this time in his ancestral village of Tan Jia Zhen, just upstream from Qutang Gorge.

> The town was wide awake. It buzzed with people rushing around, quarrelling, cautiously buying and selling. Loudly sharpening weapons. Braid cutting was in full swing. Scissors clipped off my tiny braid, so carefully cherished by my mother and my grandmother. I did not care. Let it fall to the floor. A braid on the back of a Chinaman's head is like a sign on his brow: "A Manchu slave."
>
> At the city gates, where the country roads ran up to the walls of the town, posts of soldiers were placed, armed with rifles and scissors. A soft stack of braids grew high around the soldiers' feet. The swearing of the trimmed passers-by, who cursed the revolution and covered with one hand the backs of their heads where once had been a braid, continued around the gates all day long. They would have been less ashamed had the soldiers deprived them of their trousers.
>
> The respectable intellectuals protested. "An outrage! The filthy paw of a ruffian, who only yesterday was robbing people on the road, dares to touch our noble hair. A braid is not a symbol of enslavement; a braid is a symbol of one's loyalty to the Emperor, to one's ancestors, to the great laws

> and the science of ancient times. . . ." The peasants were also indignant. "Who ever heard of such a thing? We lived for centuries with braids. Our fathers and grandfathers had braids. . . . People in the villages will laugh at us as they laugh at the tailless dogs . . ."
>
> But the coolies were glad to have free barber-shops at their service, and willingly placed their thin braids under the soldiers' scissors. They were glad to have them off. Less bother with vermin, and hair washing. They would have them cut off long ago had it not been forbidden by Manchu law.
>
> The buyers of hair were for the revolution. Never before had China brought them such a crop of cheap human hair.[1]

On October 10, 1911, a group of army officials mutinied in Wuchang, seized the city, and with the support of the Hubei provincial assembly, declared the independence of the province and the establishment of the Republic of China. Sichuan, which had resisted Beijing's attempts to nationalize the railway, announced its independence on November 27. The Sichuan Railway Protection Movement contributed significantly to the final downfall of the Qing, for the court had been forced to send troops to Sichuan to quiet the opposition to the government and had lost much of its support in Sichuan and Hubei. By December, all of southern and central China had rebelled against the imperial government. The Qing dynasty had lost the mandate of heaven.

Sun Yat-sen, a long-time leader of the revolutionary movement against the imperial government, was appointed provisional president in the newly established Republic's capital in Nanjing. Four months later, in February 1912, General Yuan Shikai, the commander of the powerful Beiyang army in the north and one of the most important political and military figures in the country, succeeded him as president of the new republic. Yuan managed to keep control over China for a few years despite an increasingly unstable political situation, but in 1915, when he attempted to declare himself emperor, provincial rulers immediately rose against him.

The first to rebel was the governor of Yunnan province, Cai E, who dispatched his troops to the central China plain to take on Yuan's forces, invading Sichuan along the way. Sichuanese commanders initially welcomed troops from Yunnan as allies against Yuan Shikai, but enthusiasm dimmed when they stayed. Hostilities broke out between the two provinces in April 1916 and resulted in a slaughter of Sichuanese civilians by Yunnanese troops, in the capital city of Chengdu. The British consul-general at the time, Meyrick Hewlett, reported about this incident:

> . . . and . . . none who were there will ever forget the massacre on the East Wall where Yunnanese seized civilians and the police at the East Gate, whom they had disarmed, pressed down their heads into the embrasures of the city wall, then one stab on the back of the neck and, dead or dying, the victim was tossed over the wall—for days the marks of blood on each embrasure told its tale.[2]

A cease-fire was declared within a week, but battles between the armies continued throughout the province for years. On the Yangtze, bands of militia and brigands thrived on the lack of order, and the escalating piracy brought river traffic almost to a halt.

Yuan abandoned his claim to the throne in March 1916 and died in June. His death ended the nominal rule of the central government and ushered in the warlord era, a time when China split up under the control of regional military leaders with large armies of their own. Military buildup in Sichuan was rapid. In 1911, there were 55,000 soldiers in Sichuan's provincial armies. By 1919, this had reached over 300,000, excluding the numerous local militia.[3]

Between 1912 and 1935, the twenty-three years from the end of the Qing dynasty to the time when the Nationalist government (then led by Chiang Kai-shek) was able to take control of the region, over 400 wars or significant military engagements were to be fought in Sichuan alone. Some were mere skirmishes between rival militia vying for the right to tax a section of the Yangtze, or a peasant army defending their land from the hungry men of a marauding army from a few counties away. Others were full-scale wars, with hundreds of thousands of troops fighting under the leadership of generals who controlled territories as large as many European countries. Some armies fought with hoes and sticks, others were armed with the most modern equipment of the times, and a few of the more enterprising of the warlords eventually acquired a handful of French and Russian planes. The comings and goings of the troops, bandits, and peasant militias, along with constant robberies and executions, became a part of day-to-day life. Tan Shihua related:

> The feet of marching soldiers echoed over the roads to Szechuan. Travelling salesmen came home with tales of the coming war, the battles that had already taken place, and districts that had been demolished. It was still fairly quiet around our town. Here the war was not between the armies, but between the police and the gangs of bandits which had increased considerably during the last few years of war and revolution. The gangs consisted of people like the nephew of the peasant Chen who lived in the bamboo forest, or of

> professional soldiers who were left with nothing to do when the war ended. . . . or of wandering farm-hands who could not find employment, and poor peasants put off their land for failure to pay their landlords. There was no lack of ammunition. The defeated armies from the North had left enough rifles in the district. Small bands of discontents would form into a larger group. They would kidnap the rich, the high officials and tax collectors, take them into the woods and demand a ransom. The rich country people began to move into the towns where they felt safer under the protection of its walls and the police force.[4]

Order of sorts was imposed again after 1916, when Sichuan was divided into about a half-dozen military districts. Originally established as garrisons with military, financial, and social responsibilities, they evolved into fiefdoms whose borders changed with the fortunes and might of the ruling generals. By the mid-1920s, there were six important warlords in Sichuan, two of whom held sections of the Three Gorges area. The four who controlled the rest of the province were Deng Xihou (known as the Crystal Monkey), who ruled the northwest of Sichuan near Qinghai, Tian Songyao (called Rotten Melon), who controlled the north, towards the border of Gansu, Liu Wenhui, who held the south from the outskirts of Chongqing to Tibet and Yunnan, and Liu Cunhou in the northeast, near Shaanxi.

The warlords Liu Xiang and Yang Sen held the Three Gorges area for years. Liu Xiang, the most powerful of all the Sichuan generals, controlled Chongqing and the western Three Gorges region, as well as the commerce of the entire upper Yangtze because of his close links with Shanghai and the Yangtze River Delta. Though he was known for his oppressive taxation policies, Liu instituted an extensive road building program, founded what was to become Chongqing University, and provided employment for thousands of peasants whose harvests had failed. General Yang Sen (Rat Face) ruled Wanxian and the eastern Three Gorges region until he was pushed northward by Liu Xiang in 1929. The most colorful of these warlords, Yang Sen was closely associated with men who were to become some of the most important figures in the history of the Chinese Communist Party. These included General Zhu De, the first commander-in-chief of the Communist People's Liberation Army (PLA), and Marshal Chen Yi, later foreign minister under Mao Zedong. Yang was an enthusiastic if heavy-handed reformer and once forced his wife, as an example of modern womanhood, to swim in the Yangtze in front of 15,000 people, a demand she complied with only when he threatened to shoot her.[5] Though much of his legacy did not endure over time, he left behind a number of modern schools, parks, and roads in the areas he had governed.

Though various administrative reforms were made at the county and city level in both the late Qing and the years after 1911, few people in the countryside were aware of them, and they had almost no impact in the rural regions of the upper Yangtze. Farmers kept on farming as they always had and tried to keep their grain and family members out of the way of the many armies and militias in need of food and men. Men without land continued to shovel coal and pull boats and knew nothing of new roads and schools. The situation was different in the cities, where questions about China's modernization and independence from foreign control were becoming more important. Though change came slowly to the interior of China, even here the sons of wealthy, educated families struggled with the same issues as their counterparts in Beijing, and wondered both how to change China and what their role would be in a society they could not yet envision.

Within a year of Yuan Shikai's death in 1916, Sichuan and Guizhou provinces to the south were once again at war with each other, both their armies conscripting local men and stopping river traffic. Far away, unbeknownst to most people in the villages of the Three Gorges, China entered World War I on the side of the Allies. Following the Allied victory, news reached Peking on May 4, 1919, that the Versailles Treaty, drawn up at the Paris Peace Talks, did not return the former German-held areas of Shandong province to China but instead gave them to Japan, which had seized them during the war. This decision prompted massive student-led protests in Beijing that spread throughout the country and resulted in nationwide strikes and boycotts of Japanese goods. Known as the May Fourth Movement, this was one of the turning points in China's twentieth-century history and the first nationwide expression of the growing resentment of the foreign presence within its borders. It was also a period of political and cultural awakening initiated by the first group of people to receive a modern education entirely unlike that of their elders, who had spent decades memorizing the classics in preparation for the traditional imperial examinations. In the county seat near the village of Tan Jia Zhen, students threw themselves into the anti-Japanese boycott. This excerpt from Tan Shihua's memoir gives a sense of the fervor that was to characterize political movements throughout the twentieth century in China.

> The town was divided into three districts, according to the number and situation of the three schools. Each school had to clean every store and storehouse in its district of Japanese merchandise. Twelve representatives of each school spent their days walking from one store to another inspecting the goods. Coolies, carrying baskets, followed them. . . . They were joyously greeted by

> the clerks of tobacco shops and by the agents of Shanghai manufacturers. These people were ready to support the student movement with all their money. The owners of small stores, trading in local merchandise, the fruit sellers, the vegetable men, the wood sellers, and the sellers of other local products, smiled at the procession. But the owners of dry-goods stores, haberdasheries, and crockery stores scowled. They thought at first that everything would go smoothly. The boys might shout awhile, perhaps take ten or twenty *da yang* contribution, and that would be the end of it all.
>
> But things went differently. For instance, a delegate entered a store. A thermos bottle and ashtray were displayed in the window. Japanese hieroglyphics were clearly written on the bottom of the pieces. The delegate raised the fragile objects and threw them on the pavement. He turned the store inside out while the owner trembled with rage. . . . Larger objects which could not have been destroyed on the spot were gaily loaded on the backs of the coolies and carried to a temple, of which the school boys had taken possession. There they were carefully guarded and, when enough things had been amassed, were carried to a distant square. Sad store-keepers and bald-headed old women cried at the delegates as they passed: "Crazy lunatics! To burn such good, expensive things—all for nothing."
>
> Those were amazing days. In Hankow, Changshi, Foochow, Shanghai, in every city in which university or high-school students were present, packages of matches and tooth powder, clocks, silk umbrellas were cast in the flames. Mirrors cracked, piles of wrapping paper and boxes of patent medicine glowed in the vengeful fires, as the members of the school committees stood around shedding tears of anger.[6]

Japan was not the only nation resented in China. Soon after World War I, dozens of English, American, French, Italian, and Japanese gunboats were patrolling the Yangtze to watch over their national commercial interests. Here, as elsewhere, tensions between Chinese and foreigners increased, occasionally resulting in serious conflicts. Hostilities among different Chinese factions also became increasingly severe. In 1919, the skipper of the American ship *Palo* reported that "the soldiers are becoming more troublesome. A Northerner caught alone by the peasants was boiled in oil. The number of executions has picked up. On one occasion, the beheading occurred on the bund opposite this vessel. The bodies were allowed to lie for several days where they fell as a horrible example."[7] During the 1921 war between Sichuan and Hubei, troops from each side, plus bandits, brought steamer traffic to a halt once more. Admiral Joseph Strauss, the U.S. Chief of Naval Operations, wrote "China is torn with strife and dissention [*sic*] . . . [which is] partly a question of [provincial] self-government and partly to undisciplined troops who get out of control and mutiny because they have not been paid. On the upper Yangtze River . . . Ichang has been looted twice within the year and Wuchang, across from Hankow, once. Steamers flying foreign flags have been fired on and have had to be escorted by river gunboats."[8]

The tension between foreign commercial interests and the well-being of local workers also intensified. In the first decades of the twentieth century, the number of men supported by working on junks or in related industries had swelled to nearly a million, between Yichang and Chongqing. The junkmen's guilds were powerful organizations which attempted, in the end unsuccessfully, to protect their members from the inroads of foreign merchants and faster, modern ships. By 1930, there were thirty-nine steamers and twenty-six motor vessels operating between Yichang and Chongqing. Together, the sixty-five ships (British, American, Chinese, Japanese, and French flag-carriers) made 879 stops in Chongqing in one year.[9] While this seems almost insignificant in comparison with the number of ships on the river today, the frequency of their turnabout, their speed, and their cargo capacity caused unemployment and widespread uncertainty about the future. To preserve the jobs of local men, foreign merchants and the local guilds came to a general agreement that certain commodities, including wood, salt, tung oil, and cotton, would be shipped only by junk. Most merchants were happy to let salt go by junk, for it corroded the iron holds of modern ships, but they were reluctant to part with the potential profit of tung oil, the most valuable export in the region apart from opium.

Tung oil is the product of the tung tree, which grows only in China, mostly in Sichuan and Hunan. Used for thousands of years in China to caulk and varnish boats, it became a key export after World War I. Between 1918 and 1937, seventy percent of all tung oil produced in China was sold to the United States, where it was used extensively as an industrial oil until the late 1930s, when newly invented petroleum-based products took its place. In the 1920s a string of eight junks could easily carry a load of tung oil worth $100,000, equivalent to approximately ten times that much today. On the local economy, the shipments were worth far more because elaborate protection schemes distributed profits far beyond the individual junk owners and their men. On each trip, the owners paid local militiamen to guard the boats against bandits, who were often in cahoots with one another, and passed the charges on to foreign merchants. Everyone along the river, except for wealthy businessmen, stood to lose if this system was disrupted.

In 1924, a British steamer, the SS *Wanliu,* defied the standing agreement and loaded a cargo of tung oil, ignoring the advice of the captain of a nearby

British gunboat, the *Cockschafer.* The move incited tremendous local antagonism. As the ship was being loaded, a riot broke out and the American merchant responsible for the cargo, a man named Hawley, rushed at the crowd. An American admiral, Kemp Tolley, described the scene in his book on gunboats on the Yangtze.

> Pandemonium soon broke loose ashore, complete with banners, shouted slogans, and speeches. Mass hysteria took hold. The mob commenced breaking up the lighters. Hawley, seeing his property in dire jeopardy, rushed ashore from *Wanliu* with more courage than good sense and began to lay into the mob, which held off from what must have looked to be a madman. Whitehorn [the commander of the *Cockschafer*], unaware that Hawley was the focus of the action, or even that he was ashore, fired a 3-inch blank, hoping to save the lighters. At this, the rioters momentarily drew back, but soon discovered the hoax and recovering their courage, went after Hawley.
>
> *Wanliu*'s master shouted to Whitehorn that Hawley was being beaten up ashore. Landing party to the rescue! By the time they arrived, Hawley was so badly battered that he soon died aboard *Cockschafer.*[10]

In retaliation for this attack, the commander of the *Cockschafer* threatened to fire upon Wanxian if the city authorities did not hand over the leaders of the junkmen's guild, considered responsible for the merchant's death. Two men, possibly leaders of the guild, just as likely not, were duly produced and immediately executed by the British, an event that further incensed local and national opinion against all foreigners.

Two years later, an even more serious altercation, remembered as the Wanxian Incident, took place between the British and General Yang Sen. At that time, Yang Sen did not have enough Chinese steamships at his disposal to transport troops back and forth to the constant conflicts along the river. He therefore commandeered foreign ships to move his men. British and American vessels were generally uncooperative, so Yang Sen had his soldiers embark dressed as ordinary passengers, draw their weapons, and load the troops while threatening to shoot the crew and passengers. On August 29, 1926, when Chinese troops boarded the British SS *Wanliu* (the same ship involved in the tung oil incident), its crew fought off the hijackers and the ship fled through a mass of sampans. Here British and Chinese versions of the incident diverge. The British claimed that one sampan sank with no loss of life, while Chinese reports stated that several sampans and other small boats were lost and that fifty-six men drowned. A large shipment of guns and bullets and 85,000 silver dollars, all belonging to Yang Sen, were apparently also lost. The next day, in Wanxian harbor, the *Wan-*

liu was recaptured by Chinese troops but freed shortly after by the British. Chinese troops then boarded two other ships that they held for reparations or the return of the *Wanliu.* As negotiations began between General Yang Sen and the British, 20,000 of his soldiers were deployed to the city of Wanxian. The standoff ended on September 5 when the *Kiawo,* a ship commandeered by the British at Yichang, freed the English ships after hours of fighting and heavy artillery bombardment of the city. Casualties were high, with about twenty Chinese and thirty British soldiers dead by the end of the fight and, according to Chinese records, 604 Chinese civilians killed and 398 injured, many of whom had gathered on shore to watch the fight. A thousand houses were destroyed.

The Wanxian Incident was largely treated as an example of Chinese insolence in Western histories of the time. In China, not surprisingly, it was perceived as an unwarranted military action and a serious affront, and the significant loss of life resulted in a major upheaval in the region. Zhu De and Chen Yi (later to become two of the Ten Marshals, China's highest ranking military officers), who had been assigned to work with Yang Sen by Li Dazhao, a co-founder of the Chinese Communist Party in 1921, were influential in launching anti-imperialist protests in the cities and towns of the upper Yangtze after 1926. The revenue of the British-run Asiatic Petroleum Corporation dropped ten percent in 1927, the result of a well-organized and effective boycott in the wake of the Wanxian Incident. Teams of Communist students were sent in from Beijing and other cities to organize study sessions in Zhongxian, Fuling, Fengdu, Wushan, and other counties and towns. Despite this, the soccer matches occasionally arranged between English and American gunboaters and Chinese students continued, and such direct personal contact was remembered with pleasure, unconnected to the actions or policies of the participants' governments.

The situation in cities throughout China became more complex as the Communist and Nationalist parties grew stronger and tensions between them increased. Sun Yat-sen remained in the political wilderness until the 1920s when his Nationalist Party (the Kuomintang, or KMT), based in Canton, began to receive Soviet aid and entered into an alliance in 1924 with the much smaller but Soviet-funded Chinese Communist Party (CCP). The Chinese Communist Party was founded in 1921 and had a cooperative relationship with the Nationalist Party, which during those years was strongly influenced by the Soviet-dominated Comintern. However, after Sun's death in 1925, tensions between the two groups grew worse. Relations between the CCP and KMT deteriorated further after Chiang Kai-shek, then the head of the Whampoa Military Academy in Guangzhou, assumed leadership of the

Kuomintang and launched the Northern Expedition, a military campaign to bring the northern warlords under KMT control and unify China. In late March of 1927, Communist-leaning city workers seized control of Shanghai, holding it until April 12, when members of the Green Gang, a notorious organized crime group with strong links to the KMT, killed several hundred of the workers and turned the city over to Chiang Kai-shek.

In the interim between these two events in Shanghai, on March 31, 1927, the Great Anti-British Alliance, a loosely formed organization including Communist sympathizers and activists, held a large rally in Chongqing against Chiang Kai-shek, warlords, and imperialism. Unaware that the crowd of several hundred thousand students and workers had been infiltrated by troops loyal to General Liu Xiang (a pragmatist now supporting Chiang Kai-shek), no one foresaw any particular danger. Towards the beginning of the rally, Yang Angong, a young Communist leader, and several other leftists walked onto the podium and were shot and killed by Liu's soldiers. In the confusion that followed, some two hundred people were trampled to death and a thousand were injured. Liu Xiang and other Sichuan warlords immediately followed up with an energetic joint campaign against all Communist activity in the region, temporarily wiping it out in Chongqing, Fuling, Wanxian, and most places in-between.

Throughout Sichuan, each of the warlord-controlled regions had departments responsible for finance, education, public works, and other functions, but civil administration was weak and the most resourceful sections of the military governments were those responsible for devising and imposing a wide range of taxes that made it possible to get the funds and supplies to maintain the armies. Taxes were levied on salt, opium, land, alcohol, tobacco, sugar, rice, pigs, paper, oil, and a multitude of other commodities going up and down the river. Local authorities imposed "winter defense fund taxes" and "bullet taxes." Farmers who refused to grow required crops such as opium instead had to pay a "lazy tax."[11] In 1934, the one year that the Nationalist government legalized the sale of opium before public opinion forced a change of policy, the KMT government earned over forty million yuan in taxes on opium transshipments. That same year, to give an example of typical sources of revenue, 19 percent of the income for the army of General Liu Xiang, who ruled the Three Gorges area, came from grain taxes, 12 percent from opium, 11 percent from salt, and another 18 percent from miscellaneous taxes of the sort described above.[12] Liu frequently levied the grain tax in advance, at one point collecting funds that were purportedly to cover the next thirty-one years, from 1931 to 1962, though the tax collectors still came back the following year.

People who grew up in this area in the early 1930s recall mostly the chaos without any clear sense of who the generals and armies were or why they were there. "There were Communists here and Kuomintang there, and other armies everywhere, and half the time the enlisted men weren't even sure which army they belonged to from month to month," explained a native of one of the inland villages, who now lives in the United States. "Even if they did, a lot of men joined one army and then another depending on who was in the region that season or who could pay or would leave you alone. People had to survive." Yang Sen is still remembered in the heart of the gorges as a man not good or bad but *youyisi*—interesting, a little peculiar, but like all warlords, out for himself and his own interests, first and last.

Beset by nearly continuous local and national wars as regional commanders and political groups vied for power, the provincial economy fell to pieces. Warlords and local generals taxed everyone and everything to raise money for their armies, and by the mid-1930s, as much as 75 percent of the arable land was used for opium cultivation. The armies were huge. In 1932, Liu Xiang and Liu Wenhui each had about 100,000 men, Deng and Tian, about 40,000, and Yang Sen, 30,000.[13] Their armies lacked arms, smoked opium, and were poorly paid. Graham Peck, an American journalist in China in the 1930s, described a Sichuanese army:

> They swarmed along the road and through the adjoining country without any sort of order. Some wore uniform coats or hats, but the remainder of their costumes was dictated by individual tastes and means. All were armed with the traditional umbrellas and carried as well numerous washbowls, tea kettles, flashlights, towels, uncooked vegetables, and extra sandals—these were slung over their shoulders or tied to their clothes with twine. Many who had guns swung their smaller effects in cloths tied to the gun barrels in Dick Whittington style. Those who could afford to rode in sedan-chairs or rickshaws, and those who had pets—dogs, birds, monkeys—carried them or led them on strings.[14]

In addition to the warlord armies, local militia imposed taxes that could change from week to week. Thousands of pirates and thieves, mostly trackers and unemployed junkmen who had lost their jobs because of the advent of the steamship, roamed the upper Yangtze, sometimes holding ships hostage until exorbitant fees were paid. Unless they directly crossed a local warlord or intruded on foreign interests, the business of robbery, extortion, and kidnapping continued with relatively little interference from any sort of authority. This period was also notable for the numerous uprisings by quasi-religious groups who thought they had special powers. These included the Red Turbans, who usually attacked local tax collection offices, and the Celestial Soldiers, also known as the Joss Troops. Like the soldiers of the Taiping Heavenly Kingdom before them who had overrun central China in the mid-nineteenth century, they believed that they were invulnerable to bullets, which they were not. In the late 1920s and early 1930s, Liu Xiang occasionally sent troops to bring these fanatical groups under control as they interfered with foreign shipping and were a menace to regular army units. Of more concern to regional military than this haphazard disorder were the organized protests and peasant movements that sprang up in the mid-1920s.

In 1928, Chiang Kai-shek's Northern Expedition to unite most of north and east China under his military control came to a successful close. With this accomplished, the Nationalist government issued a statement calling for a series of reforms and the integration of Sichuan into the republic. The Sichuan warlords ignored this, refusing to let KMT soldiers into the province and dismissing Chiang's call to reduce the troop strength of regional armies. Chiang, in an attempt to obtain further support from Liu Xiang, awarded him the title of Upper Yangtze Bandit Suppression Director and gave him large sums of money in return for a promise of troops for a KMT incursion against Hubei. Not surprisingly, Liu Xiang's men never left Sichuan, and neither did Chiang's money.

By the end of 1932, a conflict between Liu Xiang and Liu Wenhui had escalated into a full-scale war, the War of the Two Lius. While they were occupied fighting one another, the Communist Fourth Front Red Army crossed the border from Shaanxi and entered northern Sichuan on December 12, 1932. The 9,000 troops, down from 16,000 when they left their former base at the Shaanxi-Hubei-Anhui Soviet, set up a new base of operations in northern Sichuan about 400 miles from the Yangtze.[15] The Communists set about organizing agricultural cooperatives, outlawed the production of opium, and began political education campaigns. Despite the fact that local peasants sometimes believed that the Communist leaders were reincarnations of famous rebels and endowed with special powers, or perhaps because of this superstition, the new soviet had between 80,000 and 100,000 troops two years later.[16] In October 1933, Suiting, a town midway between Chengdu and the Yangtze, fell to the Communists.

For the first time, General Liu Xiang perceived the Communists as a direct threat to the region. He and Yang Sen and other military leaders mobilized more than 100,000 men in a coordinated attack on the Communist forces, driving them northward, but not out of Sichuan. The disruption to agriculture and daily life, caused by the fighting, led to a rapid deterioration of conditions in the former Soviet areas and resulted in widespread grain and salt shortages. Though the Japanese had annexed Manchuria and set up a puppet regime in 1931, Chiang Kai-shek's attention remained focused, as it would be for most of the next two decades, on the internal threat of Communism rather than on the external invaders. In October 1933, Chiang Kai-shek initiated another campaign against the most important Communist stronghold, located in Jiangxi province. Besieged by 750,000 men, the CCP decided to abandon the Jiangxi base, and in October 1934, 86,000 people set out on what was to become known as the Long March, a two-year, 5,000-mile odyssey that ended up in Yan'an in Shaanxi province, which became the base of Communist operations until their victory in 1949. Approximately 7,000 of those who began the journey reached their final destination.

By the early 1930s, the only central government institutions functioning in Sichuan were the Bank of China, the Salt Inspectorate, the Post and Telegraph Administration, and the Maritime Customs. Legal matters were handled at the district level, with the military headquarters of each garrison acting as the highest judicial body in the area. In the mid-1930s, Sichuan was in serious economic and social decline, due in part to the heavy taxes levied to support the war of 1932–1933 and the anti-Communist campaigns. Additional paper money was printed to pay troops but quickly became worthless. In 1932 Liu Xiang's army (which used the Nationalist government designation of the Twenty-first National Army) began selling large bond issues. Taxes on commodities and land increased significantly after 1933. The exorbitant taxes on goods led to high prices and a drop in demand from the outside. Salt revenues went down and more people lost jobs in this industry. The international depression led to a decrease in the demand for tung oil, still the main export of the area, and the price eventually collapsed entirely. Natural disasters also took a terrible toll. The province nonetheless remained financially independent, a reflection more on the state of the rest of the country than on Sichuan's own stability. Though it did not provide much in the way of benefits to the people paying them, the revenue from taxes in Sichuan in 1934 equaled in dollar value one-third of the total income of the entire region of China under Nationalist government rule at that time.[17]

By 1934 the growing Communist insurgency had weakened the control of the warlord generals and contributed to ever-worsening social and economic conditions, causing Liu Xiang to take the unusual step of leaving Sichuan for the first time in his life. In November 1935 he traveled to Nanjing to discuss the growing Communist threat with Chiang Kai-shek and agreed to accept a strengthening of KMT authority in exchange for military aid and supplies. As

Mao Zedong's Long March forces approached southwest Sichuan in 1935, Liu allowed Chiang Kai-shek to send 20,000 troops into the province and to establish a new provincial government in Chongqing. Though Liu retained his position as the military leader of Chongqing, along with his troops and titles, the other five warlords were required to surrender their regional commands to the Nationalist army. The new regime implemented financial reforms, instituting a wide range of new taxes. These provided fresh opportunities for graft and corruption, and another source of illicit income for Liu Xiang and numerous KMT officials.

Liu and the Nationalists ruled most of Sichuan more or less jointly throughout the mid-1930s, a period marked by a rapid succession of floods, droughts, and harvest failures. Relief funds sent by Nanjing were diverted by both KMT and warlord supporters. In 1935, 1936, and 1937, months of rain and floods followed by droughts destroyed the harvest each year. As the county histories from the region note over and over, people ate roots and grass, fled in search of food elsewhere, and starved. In Chongqing, families ate the bark off the ornamental trees in parks and special crematoria had to be built to bury the thousands of famine victims. In 1935 Chiang Kai-shek moved a stronger military force into Sichuan with the intent of finally consolidating his authority over the province, and Liu Xiang relinquished his command to the Nationalists. In July of 1937, six years after Japan had occupied Manchuria, war finally erupted between China and Japan. With this national emergency, Liu Xiang was able to regain control of his army, but he died less than a year later (some say he was murdered by Chiang Kai-shek) and his army was reabsorbed by the Nationalists. A decade later, in 1949, his rival warlords, Liu Wenhui and Deng Xihou, defected to the Communists while Yang Sen, after some hesitation, fled to Taiwan.

As northern China and then Shanghai and the Yangtze river delta fell to the Japanese, the Nationalist government, along with supporters and refugees, moved inland. With the approach of the Japanese, the Nationalist capital in Nanjing was abandoned in December 1938 and the KMT leadership and their supporters began the arduous retreat to Chongqing, a thousand miles upriver. Accompanied by hundreds of thousands of people, including entire universities, factories, and myriad other organizations, the KMT made their way by large and small boats along the Yangtze to the interior. The Japanese occupied Yichang in 1939 and bombed Chongqing and the gorges relentlessly from then until 1941, but their armies were never able to penetrate farther than the mouth of the Xiling Gorge.

From this time on, the wars of the mid-twentieth century—the Anti-Japanese war, World War II, and the Chinese Civil War, flowed into one another along the shores of the Yangtze without letup. Military conscription and political divisions forced local people from their homes. Simultaneously, hundreds of thousands of people from the outside, whom no one along the Yangtze had ever expected or wanted to see (a feeling that was completely mutual), came through and sometimes stayed.

In *Thunder Out of China*, Theodore White described the tremendous influx of people into Chongqing.

> Chungking, the refugees and exiles decided almost instantly, was a horrid place, and one of the worst things about it was the people. The downriver folk who had come up the Yangtze with their government regarded the Szechwanese as a curious species of second-grade inhabitant. It was true that rich Chungking banking families like the Young brothers, belonged to the aristocracy of wealth that made the rich of all China one family. But the average Szechwanese, with his dirty white turban, his whining singsong voice, his languid manners, seemed backward even to the most backward of coastal Chinese, who, after all, had seen street cars. The natives, on their side, regarded the downriver people as interlopers and foreigners, to be mulcted, squeezed, and sneered at; they were irritated by the crowding and the rising prices . . .
>
> The one quality that foreigners, refugees, and Szechwanese shared was strangeness. Refugees from the coast were strangers both in time and space. Retreating across the face of their country, they had receded at each remove one step closer to the ancient origins of the nation out of which they had so lately lifted themselves; when they arrived at Chungking, they were in feudal times. The natives of Chungking were strangers in time alone; the new world had moved in on them, and they could not understand it.
>
> It was not only in Chongqing, but in the gorges as well, that local people collided with strange new intruders and faced the disruption of what was to be more than a decade of war.[18]

CHAPTER 6

WAR ALONG THE YANGTZE

IN NOVEMBER OF 1938, AS THE WINTER FOG MOVED IN and enclosed the gorges in a thick damp cloud, over 30,000 people and more than 90,000 tons of equipment and supplies waited on the docks, inns, and streets of Yichang, at the mouth of Xiling Gorge. Hankou had fallen to the Japanese on October 25, spurring a great exodus westward to Chongqing, China's final wartime capital. Although preparations for the relocation of China's government, schools, and industry had been underway for some time, there was chaos in Yichang as everyone battled for a spot on a steamship for the trip upriver. Finally the chairman of Minsheng, the only passenger line still in service here, stepped in and took over the allocation of space from contending government and military offices, and in the remaining month before the drop in the water level made passage through the gorges even more hazardous, he managed to clear Yichang of almost everyone and everything accompanying the Nationalists to Chongqing.

A year earlier, after the fall of Nanjing in December 1937, the government had begun to dismantle and move factories to the interior in an attempt to preserve heavy industry and defense production from the Japanese. Other major industrial plants were also relocated. In February 1938, the Yufeng Textile Mills in Gansu, one of the largest plants in all of China, shipped 8,000 tons of equipment to Hankou by rail. From there the cargo was sent on to Yichang by steamer. In late summer, after the flood season, the equipment was transferred onto 380 junks for the journey through the gorges. Of these, 120 foundered or sank, but ninety-nine were pulled out of the river and repaired, and 359 junks arrived in Chongqing by April 1939, eight months after leaving Yichang. Within a month, the factory was operating again in its new location.

By 1940, over 400 factories and 200,000 tons of equipment had been moved from the north and east to the center of the country. China's universities also retreated to the interior. Of 108 colleges and universities operating in the late 1930s, 94 moved inland or closed down entirely, yet 40,000 college students were studying again by the fall of 1939, 8,000 more than the previous year, the result of temporarily relocating many of the country's most important educational institutions to Chengdu, Chongqing, and Kunming.[1]

In the fall of 1938 and the spring of 1939, as the steamships and junks struggled up the Yangtze towards Chongqing, then still a remote walled city, the people who lived in the Three Gorges watched this parade sail by and wondered how far behind the Japanese were. No one had ever seen anything like it. Never before in memory had so many people come through here all at once, almost all of them needing something. For a short time, thousands of travelers of every imaginable background and social level, all requiring food, repairs, trackers, pilots, and places to stay, offered natives of the gorges a glimpse of the outside world beyond the usual merchants and military. They also provided a fleeting infusion to the local economy as well. The town of Sandouping in Xiling Gorge, now the site of the Three Gorges Dam, was known for a short time as "Little Wuhan" because of all the commercial activity generated by the thousands of people who stopped there on the way to Chongqing.

This strange interlude of frenzied movement through the gorges and along the packed earth pathways and roads of Sichuan came to an end with the spring, when the winter fog lifted and the Japanese began to bomb. For the next three years, from April through October, the months when visibility is good, the Japanese did not stop their bombing. Theodore White described May 3, 1939, the first night the bombs fell on Chongqing.

> On the night of the first major bombing raid in Chungking an eclipse of the moon occurred over Szechwan. According to the folklore of China, a lunar eclipse happens when the giant dog of heaven tries to devour the moon. The dog can be prevented from swallowing the moon only by beating great bronze gongs to frighten him from his celestial meal. All through the night, between the raids of the third and fourth of May, the beating of the gongs that were rescuing the moon echoed within the city wall, mingling with the sound of fire and the many-tongued sorrow of the stricken.[2]

Though Chongqing was the main target of the Japanese bombs, other cities in the gorges and the interior were hit as well, sometimes incidentally, more often intentionally. Zhongxian was bombed half a dozen times or so each summer, with a number of casualties. Two hundred and seventy people were killed in 1940. Wushan, the only large city between Japanese-occupied territory in Xiling Gorge and the Chinese-held Wu and Qutang Gorges, was the second most important target after Chongqing because of the concentration of Chinese military forces here. It was bombed 802 times, and much of the city center was destroyed. The outskirts of the town of Dachang, north of Wushan on the Daning River, and Wuxi, a small industrial city farther north on the Daning, were also hit. Over 900 people were killed in Fengjie. While Chongqing residents carved thousands of bomb shelters out of the steep hills of the city, many of which are now used as grocery stores, hostels, restaurants, and discos, people in the gorges fled into limestone caves for protection. Sailing through Qutang Gorge, older men still point to the cave in Bellows Gorge, near the ancient Ba coffins, or across the river to another cave now filled with pool tables, and recall that it was here that their families found shelter from the Japanese.

In June 1940, Yichang was occupied by the Japanese, and the Three Gorges region became a war front. The battle that led to the fall of the city involved an estimated eighty-one Nationalist battalions with 350,000 men and was the last large-scale offensive that the Nationalists mounted against Japan during the war. In the spring, Nationalist forces began preparations for a summer offensive against the Japanese north of Wuhan near the Yangtze, but in May the Japanese struck to head this off. Planning first to weaken Chinese troops and then to corner them at the Han River near Yichang, the Japanese engaged in vicious fighting in the foothills of the Dahong mountains to the northeast of Yichang. Zhang Zizhong, the commander of the Chinese Thirty-third Army group, refused to leave the front lines and was killed by machine-gun fire, becoming the highest ranking Nationalist officer in the history of the Sino-Japanese war to die in command of a battle. After a number of other bloody engagements in late May and early June, the city fell to the Japanese on June 12. The Japanese recorded 10,500 casualties among their troops and estimated 60,000 Chinese deaths and injuries.

Yichang, described by Dick Wilson, the author of *When Tigers Fight*, as a "perilous and man-draining outpost"[3] of strategic value to the Japanese only if the Nationalists attempted to recapture Wuhan, was only 240 miles from Chongqing. The Japanese occupied the city until their defeat in 1945, using it as a base for daily bombing raids along the Yangtze and dispatching patrol parties into the mountains. Chinese soldiers later recalled that the Japanese were sometimes so close they could see the shimmering metal of their helmets. Despite such forays, Japan was never able to mount a land or river invasion beyond Xiling Gorge and, as a result, few Chinese civilians here came into direct contact with the Japanese and did not experience the direct brutality or acts of revenge suffered farther downriver and in northern China.

Though Yichang was spared the bombing that Chongqing and other cities endured, it was first half-destroyed during the Japanese invasion and then subject to a five-year occupation. Until the late 1930s, Yichang had been an active port city providing the services needed by ships and people entering and returning from the gorges. The city had withstood hard times after the turn of the century, having been sacked and looted twice in 1920 by bandit-soldiers from Badong (where Kou Zhun had tried to revive the economy a thousand years earlier), but had managed to retain a traditional flair while becoming a hub of modern commerce. At the beginning of the war years, G. R. G. Worcester, the great authority on junks and shipping on the Yangtze, wrote, "I Chang, with all its squalor, its aloofness, and its proud self-sufficiency, is one of the most characteristically Chinese cities in the whole country. The variety of its scenery, the backwardness of its inhabitants, its ancient and unchanged customs and carefully guarded traditions, the interest of its boats and the courage of the junkmen, all make it a fascinating centre for anyone interested in the age-old culture of China."[4] Kerosene, copper ingots, cotton yarn, opium, galvanized iron, and hundreds of other items from all over the world came through here, and the offices of Standard Oil and other international companies stood on the main street. Though some services, such as mail delivery, continued with little interruption through the Japanese occupation, normal long-distance commercial activity throughout the gorges almost ceased. Because of the lack of work, junk owners who normally would never have left their section of the river sailed great distances in search of employment. In the mid-1940s, boats designed for the rapids of the gorges could be spotted as far away as Shanghai harbor.

The war also brought an end to the hundred-year-old gunboat era. By the autumn of 1941 only a handful of gunboats remained in China, among them the American SS *Tutuila*, which spent the war years docked in Chongqing, and the American SS *Wake* and the British HMS *Peterel.* Both ships were in Shanghai in December 1941, when Japan bombed Pearl Harbor. Japanese officers boarded them in the early dawn. With the *Wake*'s captain ashore, the Japanese took over the ship on the spot and renamed it the HIJMS *Tatara* before his return. The ship sailed under the Japanese flag until 1945, when it was turned over to the Nationalists and became the RCS *Tai Yuan.* Captured by the Communists in 1949, it was still afloat thirty years later. The *Peterel* did not fare so well. When the Japanese military boarded and demanded immediate surrender, which was contrary to international law, the British captain refused and ordered the Japanese off his ship. They left, politely inviting him to visit their ship in return, then they fired on the *Peterel* and watched it sink. Six men were hit and killed or drowned.

The United States had supported China morally, and to a smaller degree materially, since the Japanese invasion. After Pearl Harbor, the United States and Britain joined China in the war effort as military allies. In the summer of 1941, American volunteers arrived in Kunming, in the southern province of Yunnan, to form the Flying Tigers, pilots who flew the "hump" over the Himalayan mountains to bring supplies to China from Burma after all land routes were cut off by the Japanese. In early 1942, U.S. General Joseph Stilwell was sent to Chongqing as head of the U.S. Forces in China, India, and Burma and as Chiang Kai-shek's chief of staff, with the task of encouraging Chiang to take a stronger stand against the Japanese. The relationship between the two men was antagonistic and became increasingly bitter as Stilwell pressed Chiang to reform his army and commit more strength toward defeating the Japanese rather than the Communists. Chiang in turn grew more frustrated by the interference and arrogance of someone he saw as an unnecessary interloper, and Stilwell was recalled in 1944.

In the early 1940s, while the Kuomintang was based in Chongqing, the Chinese Communist Party, whose membership had grown to 800,000, remained headquartered in Yan'an in Shaanxi province, gradually expanding its influence north of the Yellow River and in scattered "liberated areas" in the center of the country. In 1941, the Nationalist and Communist United Front against the Japanese, established in 1937, fell apart on a national level. Both armies continued to fight against the Japanese, but it was rarely a joint effort. In the Three Gorges, the 1940s were a period of tremendous political instability. Communist and Nationalist military and administrative personnel sometimes worked together, at other times, often every few months, one party or the other would seize control of the town or county governments.

In Wushan, the city in the gorges hardest hit by the Japanese, residents fled to the countryside to reduce their chances of being killed by a bomb, and schools and government offices were relocated in scattered villages. Many people spent close to four years dispersed in the countryside, returning to the city only briefly in the winter months, when the bombing ceased, to tend to their disrupted affairs. During these difficult years, Communist organizers worked hard to develop sympathy among the remaining population. Communist cadres were billeted with families still in town, and in 1941 the CCP briefly gained administrative control of the city. In order to raise much-needed revenue, the Communists instituted a tax on chickens, collecting 74,993 yuan in a year. To raise morale and increase the sense of community, they organized track and field races, social dances, and academic programs. With little else in the way of entertainment, these were extremely popular, and in sharp contrast to the austerity of life imposed by the Communists in decades to come. Once the Japanese bombing stopped after 1941 (when the Japanese concentrated their air strength in the Pacific) and the townspeople returned home on a year-round basis, the Kuomintang regained control of Wushan. In the early 1940s, the Communists had less influence there and in most other towns between Yichang and Chongqing, though a cooperative relationship continued to some extent between the KMT and CCP. In 1944, under their joint leadership, agricultural groups were formed in Wushan, Fengjie, Fuling, Zhongxian, and other counties. Despite the war, construction of public parks and athletic facilities began in various cities and new schools opened, an odd and brief mixture of improvements in the larger towns while conditions in the countryside deteriorated.

During the early war years, between 1939 and 1943, the part of China under Nationalist control, which was mainly in the southwest and central provinces (known as Free China as opposed to Japanese-occupied China), experienced an economic boom of sorts. Coal output doubled, electrical output increased sevenfold, and over a thousand miles of railroad track were laid, yet in the rural areas of the Three Gorges region most people remember the time between 1938 and 1949 as one long terrible stretch. Throughout the war years, the same cycle of ruinous weather and economic oppression continued. The Wushan and Zhongxian county histories record that all the seedlings rotted and the harvest failed in 1942 when it rained for seventy-two days straight from May to July. In 1944, there was great misery in Wushan, Dachang, Damiao, Liangping, and Fengjie, then devastating droughts in 1945 and 1948. Inflation soared wildly, reaching an average of 230 percent per year, nationwide, between 1940 and 1944. In Chongqing, prices in 1944 were an estimated 2,000 times what they had been in 1939.

Peasants still grew their own food, which should have resulted in better conditions for them and nearby towns, but they were often impoverished by the ceaseless government demands for grain. With cash worthless because of inflation, grain became the medium of financial exchange, including the payment of taxes. Peasants were required to sell their grain to the state at fixed prices and have it delivered at their own expense to sometimes distant transport stations. Although the Kuomintang passed a law in 1930 that restricted rents to three-eighths of the annual value of the tenant's main crop, this was never implemented; landlords set rents at will and charged interest on overdue payments at an average of 30 to 60 percent interest.

Practically everywhere, the peasants and other ordinary people were tired of the chaos of war. Girls and women were sold off for money to feed the remaining members of their families, and now, instead of the warlord armies, the farmers had the Kuomintang military to deal with, which by most accounts was not an improvement. Conscription into the Kuomintang army was particularly harsh in the mid- and late-1940s in Sichuan and along the Yangtze. Men in villages tell of how the folly of an afternoon trip to the county seat to deliver grain ended with them in the army for the next two years, or of being forcibly marched off from the fields or while talking to a cousin in the marketplace. In theory all men aged eighteen to forty-five were subject to the draft, but almost everyone who could afford to pay his way out of service did so. Theodore White described in *Thunder Out of China* how a black-market recruit was trussed and bound by a press gang and sold for "the equivalent of the purchase price of five sacks of white rice or three pigs."

> So many bought their way out of the draft that village heads could not meet their quotas; in order to supply the requisite units of human flesh, organized bands of racketeers prowled the roads to kidnap wayfarers for sale to village chieftains. Army officials engaged in the traffic on their own, and they made no protest no matter how decrepit the recruits' health. . . . In one Szechuan district the village headman stationed himself at a crossroads with armed soldiers and seized a fifty-year-old man and his grandson. The boy was leading the grandfather to the hospital, but it made no difference; off they went to the recruit camp. In two instances village chiefs took their gendarmes to a river to seize boatmen. The boatmen produced cards proving they were engaged in an essential occupation and were draft-exempt. Two were drowned; two were beaten to death; the fingers of another were cut off; more than ten were drafted. . . . Able-bodied men deserted their villages and formed bandit gangs in the hills to wait until the drive was over. Peasant youths refused to haul pigs and rice to city markets for fear of being seized on the road.[5]

When Japanese Emperor Hirohito surrendered on August 15, 1945, following the American bombing of Hiroshima and Nagasaki, there were massive celebrations throughout China. In Wushan, local histories describe thousands of people winding through the city streets as part of a torchlight march to celebrate the victory. Even in the midst of their shared delight and relief, the eternal dispute between local Communist and KMT factions erupted, foreshadowing the total breakdown of relations between the two parties once the common goal of getting rid of the Japanese was gone, and the festivities were disbanded by the county head who saw a victory march about to turn into a street brawl. In 1946, CCP troops were brought in to defend the county office from the Nationalist army, and special units were sent into the countryside to protect organizers, setting up cooperative farming teams as part of an early collectivization program opposed by the Nationalists. Despite the tensions, there was some sense of increased stability and a general expectation in the Yangtze towns that, tired as they were of chaos and corruption, the end of war with Japan might bring an opportunity to rebuild, presumably under the Nationalist leadership that continued to dominate.

After the Japanese defeat, however, conditions for the Nationalist troops in this area were still, by all accounts, awful. Men were often half-starved, inadequately clothed, and subject to violent discipline. Desertion was widespread. During a march from the north of China to the southwest, one division lost 3,000 of its 7,000 men while transiting Sichuan, the province from which most of the soldiers had come. Entire companies simply went home. After Chiang Kai-shek led the Nationalist government into retreat in 1949, thousands of men from the Three Gorges region who had been swept into the Kuomintang army during the 1940s ended up in Taiwan. Few knew where they were going; none expected to stay for long. Mostly peasants with little education or money, they had no opportunity to communicate with their families before leaving the mainland, and once on Taiwan, life was not easy as they grew older, often alone and with few resources.

When the rules on visits by residents of Taiwan to China were relaxed in the mid-1980s, first for family reunification and later for tourism, elderly men streamed into Chongqing and fanned out all along the river to visit wives they had not seen in forty years, children with grandchildren of their own, and unknown brothers and sisters. Filling entire planes, hundreds of ex-soldiers returned together, bringing televisions, electric rice cookers, and brightly colored fabrics, not knowing what to expect or how they would be welcomed. For some returnees, this was to be the only visit. Travel was expensive and it could be difficult to resume relationships severed for half a century. The burden of expectations from relatives in the countryside, poor farmers who had heard in recent years about Taiwan's wealth, could also be heavy.

Some men from Taiwan made substantial contributions to their home villages, and many had already found ways to send money back over the years, but others could not or did not, and their original families often found it hard to accept that husbands and fathers who had spent decades away could return so changed but little better off. For others, there were happier stories, men who never remarried or were widowed in Taiwan, who came back to a wife and children, and sometimes parents still alive, who welcomed them and with whom real bonds survived or developed. Whatever they found, and however simple their lives in Taiwan, the old soldiers had been gone so long that few could imagine a permanent return to the hardships or confinement of rural Sichuan life. Ties to the *lao jia*, their original home, are strong though, and many men return repeatedly, sometimes for months at a time. Nonetheless, the feelings of conflict, of belonging and not belonging, are almost always there, if difficult to articulate for men who come from peasant families and are not used to discussing such things.

In the mid-1990s, several dozen elderly ex-KMT soldiers, a few Hong Kong and Western business people, and I were waiting for a plane to Hong Kong in the new wing of the Chongqing airport. We had been informed that a plane had arrived but would not be departing in the foreseeable future, and we were left to wait. An hour later, the passengers began to grumble. It was past lunch time and there was no food. Like stage props, one shop sign proclaimed "Snacks and Noodles," another "Gifts and Candy," but they stood empty. You could buy a rug but nothing to eat. In time, an attendant appeared with a rolling cart full of cans of orange soda lined up next to a flight manifest. She made her rounds, inspected each passenger's boarding card, and then issued one can per person and asked for a signed receipt.

Some of the foreign business people were amused by this cumbersome process, almost nostalgic that the China they had known in 1979 lived on somewhere, but it brought back no such fond memories for the former soldiers from Taiwan. They began to mutter. What was this in China that you couldn't buy a bowl of noodles and had to sign your name for a can of soda? An airline employee then announced that there would be lunch after all, but because there was not enough to go around, it would be only for first-class ticket-holders. This piece of news, taken as full of political implication, caused a near riot. "They say this is the new China, where everyone is equal," one old man shouted, "yet only those in first class eat!" "What has happened here the past forty years?" another man added, "Nothing! In Taiwan, everyone eats. We come back to China after forty years and we're still not good enough to eat."

Anger on Chinese streets is often sudden and extreme, so the level to which their irritation escalated was not surprising, but the interaction between these old men who wanted lunch and the airline attendants who had never left China reminded everyone all too well of the unresolved conflicts of the past half-century. The airline personnel had not expected much of a reaction to the news about the delay or lack of food. They were used to telling people that there was no food, no heat, no telephone, no whatever, and expected the passengers to be accustomed to it as well. Generally, despite recent improvements, domestic passengers were. The dynamics here were different. The age of the men alone made it impossible to ignore them for one could not completely disregard the fury of a group of eighty-year-olds. The customary regard for the elderly and a concern that someone would drop dead in rage required a different response. These former soldiers were also visitors from Taiwan, and it was necessary to make a good impression, yet they were clearly local people, men with familiar thick and unintelligible accents that identified them as peasants from nearby hinterland villages. In the minds of the city workers, these men were still backward locals, whom one would not normally encounter on a plane or pay much heed to, but here they had to. As the murmur of protest turned into a roar of vituperative complaints about the airport in particular and China in general, boxed meals were found for everyone in economy class.

Cardboard lunch boxes in hand, the ex-soldiers and all the rest of us untied the pink twine that held them closed. We ate our boiled eggs and gritty yellow cookies and drank our small cans of pineapple juice. The men complained a bit because the food was cold, but were momentarily satisfied that they, peasants who had been treated horribly here decades ago, had won the lunch battle. The plane was eventually repaired, or the pilot found, and in the dimming light we flew off to Hong Kong. There, clutching bags of herbs and souvenirs and photos, the men disembarked and headed off to rebook their missed flights home.

In the midst of World War II, while China and Japan remained in a tense stalemate in the Three Gorges and the Americans fought against the Japanese in the Pacific, all three countries were busy making plans for a dam in the gorges. All saw hydroelectric power as the key to China's future, and each expected to benefit by it. Soon after the Japanese occupied Yichang in 1940, confident of the future, they made a detailed aerial survey of the entire Three Gorges area and began planning for a high dam that would generate 30 million kilowatts of energy. Known as the Otani plan, after Otani Kozui, the engineer who designed it, the dam was to be 591 feet high and 6,562 feet long, with a hydropower plant and multistage shipping locks or an elevator capable of lifting a 2,000-ton ship. Otani's comprehensive plan came to naught, but the topographic surveys were intercepted by Chinese intelligence and used by Chinese and American engineers when they began designing their own dam two years later.

Like Sun Yat-sen, Chiang Kai-shek saw the exploitation of energy from the Yangtze River as key to China's future prosperity and as a link with Sun's legacy. Twenty years earlier, Sun Yat-sen had hoped that electricity, and the Three Gorges Dam in particular, would make China's economic development possible. In a 1924 speech in Canton, he presented a romantic vision of the dam.

> If the water power in the Yangtze and Yellow Rivers could be utilized by the newest methods to generate electrical power, about one hundred million horsepower might be obtained. Since one horsepower is equivalent to the power of eight strong men, one hundred million horsepower would be equivalent to the power of eight hundred million man power. . . . Manpower can be used only eight hours a day, but mechanical horsepower can be used all twenty-four hours. . . . If we could utilize the water power in the Yangtze and Yellow Rivers to generate one hundred million horsepower of electrical energy, we would be putting twenty-four hundred million men to work! When that time comes, we shall have enough power to supply railways, motor cars, fertilizer factories, and all kinds of manufacturing establishments.[6]

Sun himself drew a series of seven designs for a Yangtze dam, which he showed to Chiang Kai-shek. For Chiang, Sun's dream became an integral part of his own postwar reconstruction plan. As early as the 1930s, Chiang solicited a study on ways to improve the course of the Yangtze, carried out with the assistance of American and German advisers, and gave approval for a survey of Three Gorges by a team of four engineers, three Chinese and one American. This was conducted under the supervision of the British-trained economist C. C. Chien, who also advised Chiang to develop an industrial base capable of resisting Japan. In early 1933, the group recommended the construction of a 42-foot low dam in Xiling Gorge, either at Gezhouba (later the site of the first Yangtze dam, built in the 1970s) or at nearby Huanglingmiao. The dam was designed to improve navigation and irrigation in the upper reaches and to provide 300,000 kilowatts of energy, almost a hundred times the national hydroelectric generation capacity at that time. The construction cost was estimated at 80 million silver dollars (approximately U.S. $363 million in 2001), which put its realization out of the realm of possibility.[7]

In 1944, although the war against the Japanese dragged on and cooperation with the Communists was long over, Nationalist officials turned to the idea of a great Yangtze dam as part of an overall postwar construction plan. That summer, the National Resource Commission (NRC), the division of the Nationalist government responsible for economic development, invited John

Savage, who had been the chief design engineer for the U.S. Bureau of Reclamation, to come to China. Savage had worked on all the important American dams of the time, including the Hoover and the Grand Coulee dams. He had overseen construction of the Tennessee Valley Authority (TVA) and personally designed the Norris dam, the biggest dam of that project.

The TVA, established in 1933, was a comprehensive program to control flooding, improve navigation, provide electrical power in seven states, and improve the living standards for the region's farmers. The damming of the Tennessee River and its tributaries protected thousands of acres of farmland and revitalized a large and economically depressed area of the United States. The TVA eventually generated over 12 billion kilowatts of energy per year, making it one of the largest electricity producers in the world. In the mid-1940s, China had a total hydroelectric production of 3,830 kilowatts of energy, with a total national electrical output of only 73,577 kilowatts in the region under Nationalist control.[8] Electrification was seen as the key to industrialization and successful postwar recovery, and the TVA was seen as "a model to develop all resources of a major river basin under one comprehensive program sponsored by a central government" that the Nationalist government hoped to emulate.[9]

This was the era of great dams, and in the 1940s the projects were generally deemed successes worth whatever sacrifices had been necessary, and an ideal model to export. John K. Fairbank, the preeminent scholar of modern Chinese history, wrote, "TVA makes sense in China. The use of public funds for big public works and water control, the government and the individual citizen cooperating in the application of modern technology to the ancient problems of the soil, the state helping the small man to help himself—that is the most clear-cut democratic ideal in China. In our relations with economically backward peoples, TVA is a primary asset."[10] During the Great Depression, American dam construction provided work for thousands of jobless men and transformed hundreds of thousands of acres of California and Western desert into fertile agricultural land. The Bureau of Reclamation and the Army Corps of Engineers, in collaboration with the TVA planners, expanded their flood-control projects to the basins of the Missouri, Colorado, and Arkansas rivers. American engineers and economists saw the creation of these huge dams as a solution to the economic and related social problems of vast areas of the world lacking irrigation or suffering the ravages of uncontrollable rivers. Liangwu Yin, whose Ph.D. dissertation provides a detailed and comprehensive overview of China's efforts to build a dam in the Three Gorges, described the goals of the TVA:

> .... the TVA planners were intoxicated with their initial success. They then collaborated with the Bureau of Reclamation and the Army Corps of Engineers in an attempt to duplicate the feat in other river basins in the United States: the Missouri River, the Colorado River, the Columbia River, and the Arkansas River. In the meantime, New Dealers and their supporters pushed for the TVA idea abroad. They viewed the TVA as the greatest peacetime effort at social change for oppressively poor areas, and as the closest approximation of a perfect paradise the United States had ever accomplished. This concept, in various forms, received very wide support in many underdeveloped and undeveloped nations. The TVA was looked upon as a model for the speedy realization of economic and social progress. For a while, there were advocates of a "DVA" for the Danube Valley, an "AVA" for the Amazon Valley, and a "YVA" for the Yangtze Valley.[11]

Anticipating future loans and technical assistance from the United States, China's Nationalist leaders thought this an ideal model to follow, and in November 1944 the government adopted a policy making large-scale hydroelectric projects, to be developed exclusively by the central government, one of the top national priorities. Economists drew up a five-year plan based on the Tennessee Valley Authority and called for the construction of a dam that would control floods, irrigate 12.7 million *mu* (2.1 million acres) of farmland, and generate 477,000 kilowatts of energy within this time frame.[12]

Much taken by the idea of working on a Yangtze River dam, John Savage headed to China for an inspection tour in the summer of 1944, somewhat to the distress of the Chinese authorities, as the Three Gorges were in the middle of a war zone. Savage, accompanied by General Wu Qiwei, the commander of the Yangtze Defense Corps and deputy commander of the Sixth War Zone; an American army photographer; and various other engineers and soldiers, set off by steamship from Chongqing on August 19. Near Fengjie they switched to sampans and completed the survey on foot, coming within three miles of Japanese-held territory to look at Xiling Gorge. Savage selected a site near Nanjinguan at the mouth of the gorge as his first choice. Fifty years later, in 1994, work began on the current Three Gorges Dam at Sandouping, just to the west of this spot.

Using the captured Japanese aerial topography maps, Savage spent the next few months on the outskirts of Chongqing designing the dam. In October 1944, he presented a plan to the Chinese government that would provide 10.56 million kilowatts of electricity, a storage capacity of 22 million acre feet (an acre foot equals 43,560 cubic feet), large enough to control the largest recorded flood along the Yangtze, and a reservoir of 50 million acre feet. Savage proposed five sites and two different kinds of dams. The first was a 738-foot-high straight gravity dam that would raise the water level 525 feet. This included a series of bomb-proof diversion tunnels housing 96 turbogenerators capable of generating as much energy as the Hoover, Grand Coulee, and Shasta dams combined. The alternate design for the dam, with the power plant above ground, was the same height as the first, but longer and lacking the indestructible tunnels. Savage estimated that the total cost would be less than one billion U.S. dollars, recoupable within five to seven years of operation.

The 525-foot increase in the water level—50 feet lower than the Three Gorges Dam designed in the 1990s—would have created a reservoir 250 miles long instead of the 360 miles now planned, but still would have made the Yangtze navigable by 10,000-ton ships as far as Chongqing. This dam, too, would have displaced a great number of towns, villages and historic sites. However, with the population of the region less than half of what it is now and a smaller reservoir affecting fewer households, it was not a project, given the timing, likely to raise much opposition either within China or internationally.

The U.S. Bureau of Reclamation gave an enthusiastic account ot the project:

> Tall as a skyscraper, this dam as planned will reach higher into the sky than Boulder Dam of the Colorado River, now the world's highest (750 feet [*sic*] compared with 726). Its hulking mass —a third greater than Grand Coulee dam on the Columbia River, the present title-holder (15 million cubic yards to almost 10 million)—dwarfs the largest pyramid. A record 250-mile reservoir backed up behind the dam (50-million acre feet capacity compared to Boulder's approximately 31-million) will store water for the mammoth irrigation project, to maintain navigation, for flood control, and for power production.[13]

In the fall of 1944, Donald M. Nelson, the head of the American War Production Board, was sent to China as a personal representative of President Franklin Delano Roosevelt, to look at ways to encourage economic development. Nelson quickly became a strong advocate of the dam, as did Secretary of the Interior Harold Ickes; T. V. Soong, China's foreign minister; and Chiang Kai-shek himself. The Nationalist government planned to begin building the dam immediately after the end of the war, and Nelson lobbied for U.S. support, encouraging Roosevelt's interest in the project and obtaining his approval to appoint a Yangtze River Survey Committee as the key body to work with the Chinese in what was then envisioned as a joint Sino-American project. Describing the possible benefits to the United States, Nelson wrote to

President Roosevelt, "The Yangtze development would mean large exports, the stimulation of key industries and many jobs for workers. Several agencies of our government, and members of the House and Senate, are actively interested, as are a number of private industrial and engineering concerns."[14] Roosevelt responded with enthusiasm, but after his death in April 1945, his successor, President Harry Truman, showed no interest in the dam. Nelson left the government shortly thereafter and his successors were unable to rouse the same level of support for the project.

In face of this lack of official interest, Savage asked the Bureau of Reclamation for direct assistance without the involvement of other U.S. government agencies, though such aid would still require State Department approval. Following up on this in January 1945, T. V. Soong submitted a formal request to U.S. Ambassador Patrick Hurley for the U.S. Bureau of Reclamation to assume primary responsibility for the design and construction of the Three Gorges Dam. The State Department had initially opposed this plan on the grounds that the dam would generate too much power to be used effectively by China, was not sufficiently integrated into an overall plan for national industrialization, and would require massive American financial support. Nine months later, after lengthy negotiations, on August 30, 1945, only two weeks after the unconditional surrender of Japan that ended World War II, the State Department officially informed the Department of Interior that it would give approval to the Bureau of Reclamation to build the dam in the Yangtze gorges. In November 1945 the agreement was approved and signed by the U.S. Department of State, the Department of the Interior, and the Bureau of Reclamation, and the National Resources Commission of the Republic of China. Announcing the agreement, Harold Ickes stated that engineers would " . . . prepare the designs and specifications for what may be the largest concrete dam ever built . . . to control the Yangtze and free millions of people in its valley from the scourge of floods and famine."[15] Despite earlier reservations, the U.S. government was caught up in a wave of enthusiasm for the project and tentatively agreed to give China a three-billion-dollar loan for construction of the dam. Official enthusiasm ran high in China too, with the *Zhongyang ribao* (Central Daily News), an organ of the Nationalist Party, declaring in the fall of 1945 that "the 'YVA' is the best guarantee of China's future prosperity and democracy."[16]

On December 7, 1945, the fourth anniversary of Pearl Harbor, the National Resources Commission of China sent the U.S. Bureau of Reclamation a check for $250,000 and requested that work on the dam begin immediately. In January 1946 U.S. Ambassador George C. Marshall persuaded the Nationalists and Communists to agree to a cease-fire, though it was clear that neither side was committed to it. Nonetheless, with Savage already working on the dam and anxious for the Bureau of Reclamation to get started, it seemed, for a few months, that the dam would go ahead as planned. This was not to be. The cease-fire between the KMT and the CCP collapsed in Manchuria in May 1946 when both sides rushed to seize weapons and territory from the departing Soviet troops. As Chiang Kai-shek's grip on China began to unravel, so did hopes for the Yangtze dam.

Over the next year, Communist forces captured large areas of north China and crossed the Yellow River. The U.S. Bureau of Reclamation had requested that 150 Chinese engineers arrive in Denver by June 1946 and bring with them complete hydrologic and topographic information, but by December only forty-two had reached America. It was becoming clear that in the postwar chaos, with the Chinese economy in shambles, a million Japanese to repatriate, and an all-out war with the Communists developing, there was no way to make progress on the dam. A year later, in May 1947, China and the United States suspended work by mutual agreement in light of the runaway inflation and political instability.

In March 1948, Premier Chang Chun, the president of the Executive Yuan of the Republic of China, gave a speech at a Rotary Club meeting in Nanjing on the importance of the Yangtze Valley region and the need for foreign assistance. He assured the audience that the Three Gorges project would be resumed "as soon as the situation improves."[17] The situation did not improve though, at least for the Nationalists, and by mid-1948 it was clear that the balance of power had begun to shift toward the Communists. The next year, 1949, was one of constant disruption along the Yangtze River and ended Chiang Kai-shek's hopes for a Yangtze Valley Authority and a rapid modernization of China based on the energy to be provided by the dam.

The gains and losses in Wushan county in the late 1940s were typical of the struggle between the Communists and the Nationalists throughout the Three Gorges region. In early 1949, the Communists lost much of the ground they had gained in the county, and the Nationalists imposed martial law and draconian punishments for violating the new regulations. In response, the local Communist leadership intensified its intelligence efforts to infiltrate the Nationalist ranks here. In September 1949, Su Tingyu, a Nationalist general exasperated by the widespread corruption and miserable conditions of the soldiers, led a revolt against the KMT command and signed a secret agreement with the underground CCP. In November, Communist troops occupied Yimiao, seventy miles to the north of the town of Wushan, and the Nationalists cracked down in Wushan, arresting dozens of underground Communist Party officials. This was a futile and last-gasp operation for Nationalist officers and men worn out by the years of fighting. The next month, the county fell to the Communist People's Liberation Army (PLA).

In China as a whole, the main forces of the PLA had begun to move southward in the fall of 1948. After victories in north and central China in the spring of 1949, the final battle for the Yangtze began with tens of thousands of Communist troops advancing to support the pockets of Communist militia that had been fighting against the Nationalist army for over four years. The PLA captured Nanjing in April 1949, Shanghai in May, and Yichang in mid-July. In early December, the towns in the Three Gorges were occupied by Communist forces as Chiang Kai-shek and two million of his supporters fled to Taiwan. The dam was soon a part of a dream of economic might for China's new Communist rulers and, as it had been for both Sun and Chiang, a symbol of the possibility of power over man and nature.

CHAPTER 7

REMAKING MAN AND NATURE

THE ESTABLISHMENT OF THE People's Republic of China (P.R.C.) was declared in Beijing on October 1, 1949, but it was to be several more months until the Three Gorges region came under Communist control. In October, the Second Field Army, under the command of Deng Xiaoping and Liu Bocheng, both natives of Sichuan, entered Sichuan province from Hubei in the east with the assistance of the First and Fourth Field armies. The leadership of Sichuan was to come out of these military groups for decades. The upper Yangtze fell county by county as Chinese Communist Party Second Field Army forces moved first southwest and then eastward from Chongqing, occupying Zhongxian on December 7, Wanxian on December 8, and Wushan on December 20. Yichang, in Hubei province near Xiling Gorge, had been under Communist control for several months already, since July 16, 1949, when the Fourth Field Army, led by Lin Biao, occupied the city. Kuomintang troops in the gorges, left behind when Chiang Kai-shek left for Taiwan in early December, struggled on without him for a few months, fighting a haphazard guerrilla war against the Communists, until gradually the distinction between militia groups and bandits became less clear, as has often happened in Chinese history.

Jung Chang, in her memoir *Wild Swans*, described her mother's journey through the Three Gorges in early 1950.

> The Communists had taken most of Sichuan only within the last month. It was still infested with Kuomintang troops, who had been stranded there when Chiang Kai-shek had abandoned his resistance on the mainland and fled to Taiwan. The worst moment came when a band of these Kuomintang soldiers shelled the first boat, which was carrying the ammunition. One round hit it square on. My mother was standing on deck when it blew up about a hundred yards ahead of her. It seemed as though the whole river suddenly burst into fire. Flaming chunks of timber rushed toward my mother's boat, and it looked as if there was no way they could avoid colliding with the burning wreckage. But just as a collision seemed inevitable, it floated past, missing them by inches. Nobody showed any signs of fear, or elation. They all seemed to have grown numb to death. Most of the guards on the first boat were killed.[1]

Everyone was exhausted. During the first six months of 1950, the first year that the Three Gorges were under Communist control, a series of misfortunes hit the region. Droughts, insect infestations, and floods along the upper and middle Yangtze made life miserable. Ever mindful of the traditional correlation between environmental disaster and a nonrighteous government, the new Communist leaders and the People's Liberation Army brought relief supplies and patched the dikes, but their efforts were not enough to stop the disaster. In Wushan and Fengjie counties alone, scores of people died of starvation by the fall.

Political, economic, and social reform and reorganization began quickly. In a sudden reversal from previous times, one's future was often determined by how badly he or she had once been treated, as everyone was categorized into newly defined political and economic classes. Land reform—the redistribution of land from wealthy landowners and prosperous peasants to the landless—started almost immediately, as did other changes. Thousands of primary schools were founded, providing a chance for a basic education for village children. New legal rights were instituted for women, allowing them freedom of choice in marriage and property ownership. Political campaigns targeted people who were a threat to the new state and its goals. In 1951 the Three-Antis Campaign (against corruption, waste, and bureaucracy) led to the deaths of approximately 500,000 to 800,000 people throughout China, but almost completely eliminated corruption for years to come. The Five-Antis Campaign the following year (against bribery, tax evasion, fraud, theft of state property, and theft of state economic secrets) was focused on weeding out and limiting the power of capitalists, this time punishing them more often with fines rather than prison or execution.

Foreign companies were nationalized, beginning with Shell Oil in 1951. All remaining private enterprises were nationalized in 1955. Foreign steamships had ceased to operate by the late 1940s, and Minsheng, the first Chinese-owned steamship company, founded in 1923, was taken over by the state in the early 1950s. In 1955, a campaign to uncover "hidden counter-revolutionaries," including "spies for imperialist countries and the Kuomintang, Trotskyites, ex-Kuomintang officers, and traitors against the Communists" resulted in the persecution and expulsion of thousands of Communist Party members who had had connections to the Kuomintang. From this time on, family background became of even more crucial importance.

The policies of the People's Republic reached throughout China in a way those of almost no other government had, and this personalization of politics meant that whether you lived in Beijing or Fengjie, your life was changed significantly and permanently. For centuries past, the inhabitants of the Three Gorges had struggled against natural disasters and taxes and had attempted to avoid the warring armies that periodically swept through. If they survived all this, most people went about the business of farming or studying or selling things with relatively little interference from the authorities. Ordinary families, if all went well, had little reason for contact with government officials and, particularly in remote areas of Sichuan or Hubei province, rarely had a sense of the presence of a central government or felt its impact on their day-to-day lives. They did after 1949.

I met in Boston in the late 1990s with a group of people born in towns along the upper Yangtze in the 1930s and early 1940s, all men and women who had grown up in this period of constant turmoil. Fifty years later they were living in the United States with grown children who had come to America and stayed on to become successful dentists and computer experts. On a Saturday morning, former landlords and poor peasants sat together, somewhat awkwardly, in a gymnasium while Chinese school was in session, all of them waiting for their grandchildren. Their stories are part of a collective, if fading, memory, one unique to each family or village, yet shared by all of China. Their lives reflect the tragedies and disappointments that sooner or later affected nearly everyone born in China during the first two-thirds of the twentieth century, plus some once-unimaginable events that took a few of them from remote towns along the Yangtze and brought them to suburban New England. Regardless of their positions in the Chinese political and social spectrum, the decades of political upheaval that led to societal and personal tragedy meant that almost everyone suffered along the way.

Mrs. Liao, a retired teacher, was born at the beginning of the Sino-Japanese war in a village halfway between Chongqing and Qutang Gorge and grew up during the Civil War. Her family was poor and as a child she searched the streets where people husked their grain for leftover kernels to help feed her parents and brothers. Around 1947, during the Civil War, a neighboring landlord family with only one child, a boy of about six, hired a tutor for him. Since he had no brothers or sisters to study with, they asked Mrs. Liao, who was then about nine, to join him, for she was both old enough to watch over him and quick enough to master the lessons and oversee his studies. This was how she came to read and write properly and to get an education when neither her parents nor her younger brothers did. Her education ended abruptly with the Communist victory, when the landlord family was driven out. This might have meant the end of any further studies and a return to the fields, but as one of the most educated people in the village, even though she was barely thirteen, she suddenly found herself in charge of the newly founded primary school. After teaching for six years, she went on to a teacher's training school in Fuling, and then on to a teacher's college and a long career in education.

Another former resident of the area, Mr. Wang, came from the outskirts of Yunyang, the site of the Zhang Fei temple. With the slight figure and fine features of actors depicting wealthy landlords in old Chinese movies, he came from a family of traditional Chinese doctors and as a teenager in the 1940s studied with his father and assisted with patients. During the war years, his father and uncles often provided free care for poor families, but they also owned land tilled by others that contributed to their income. With Liberation, as the founding of the People's Republic is called in China, and land reform, and few households to take land away from, his family became a target. Both his uncle and grandfather were beaten and killed, and his father decided to move the family to a bigger city where they could start over. Despite their very different experiences, both Mrs. Liao and Mr. Wang remember their youth with weariness.

Land reform was completed along the upper Yangtze by late 1952. Throughout China, hundreds of thousands of landowners were killed in the process of expropriation and redistribution. Landlords were attacked and killed in the Three Gorges as well, but it was a more peaceful and popular transition here, mainly because of the limited amount of land to be redistributed and little resistance. Over the next two years agricultural cooperatives were formed in most of the river towns. In these, peasants retained title to their land but farmed together and divided the harvest according to the amount of land and labor contributed by each family. In the winter of 1956, the cooperatives were merged into collectives, emulating the Soviet collective farm model. Private land ownership was eliminated and members were compensated according to their labor, without regard to material contributions such as land or farm tools. In Sichuan, problems related to grain production began in the mid-1950s when the state began to monopolize the grain market and demanded increasingly higher grain quotas. One of the key political figures during the first decade of Communist rule in Sichuan was Li Jingquan, an ultraleftist originally from Shaanxi province, who had been a follower of Mao Zedong since the 1930s. Private trading was banned when Li attempted to prove that the early stages of collectivization would result in more grain. As incentives for peasants to grow grain diminished because of the increasing communal structure, grain-purchasing targets increased and became more unrealistic, leading eventually to an irreconcilable tension.

As part of the movement to rid China of both foreign imperialism and superstition, Christianity also fell under suspicion. Ministers and priests were driven out, and churches closed. A large Catholic population had grown up in the towns along the upper Yangtze in the early twentieth century, and many of the Chinese Church members did not abandon their religion lightly. Wushan's Teacher Yang recounted with five decades of anger the loss of priests who had been made to abandon their vocations and were forced to marry. Liberation meant the end of religious life for thousands of parishioners. Teacher Yang, too, had been required to marry and to give up all public allegiance to Catholicism, but forty years later she was again a major force in a newly resuscitated church.

In the spring of 1956, Mao Zedong, chairman of the CCP and president of the P.R.C., launched the Hundred Flowers movement, with the slogan "Let a hundred flowers bloom, let a hundred schools of thought contend." This initially seemed to suggest a period of greater freedom for literature, the arts, and scientific research, and more opportunity for free discussion. A year later, Mao encouraged intellectuals, meaning educated urban residents, to speak openly of their dissatisfactions with the party and its officials, but this unleashed a wave of criticism far greater than what Mao had expected; when too many people spoke out, saying far too much, a crackdown followed. The Anti-Rightist movement began in 1957. Hundreds of thousands of people were arrested and sent to prison or into exile for criticizing the government. Many citizens labeled as rightists did not have this stigma formally removed until after 1978 and spent decades in jail or unable to find work. Universities, schools, and government offices were all required to ferret out and denounce 5 percent of their employees, a quota established on a national level. In Sichuan, where the zealous Li

Jingquan was responsible for carrying out the Anti-Rightist movement, the quota of arrests was set even higher.

By late 1957, a new political slogan, "more, faster, better, more economically" (*duo, kuai, hao, sheng*), was appearing on billboards and in newspapers. A harbinger of what was to come, the P.R.C. began a drastic program in 1958 to "complete the building of socialism ahead of time, and carry out the gradual transition to communism." Part of this program was to build bigger and better infrastructure projects, and this included new irrigation projects and hydropower plants. In early 1958, for the first time since the founding of the P.R.C., the government of China once more turned its attention to the possibility of a Three Gorges dam. Premier Zhou Enlai was assigned the task of overseeing new studies and preparations, but this renewed interest in the dam was set aside with China's embrace of the Great Leap Forward, an ill-fated and irrationally structured program to increase industrial production to overtake Great Britain within fifteen years, which resulted in the worst famine in China's history. It was not until 1970, with the country still in the grips of the last phase of the Cultural Revolution, that work began on the long awaited Yangtze dam.

With Mao promising that the Great Leap Forward would make it possible to double steel output and achieve the rapid growth of heavy industry, each province was given impossible quotas and the entire country turned its attention to smelting iron and steel. In the summer of 1958, all able-bodied men were required to put in long hours at newly opened smelting plants hastily constructed on the mountainsides, in addition to continuing their farm work. Entire families in cities and towns spent hours every day working at homemade neighborhood furnaces, melting down cooking pots and spoons to make the steel that would push China to rival Great Britain. For all this effort, the metal produced was of such poor quality that it was mostly useless. Thousands of men were also sent off to construction brigades to work on dam-building projects encouraged by the central government. Many of these were built under tremendous time pressures, often with poor quality materials by people with no knowledge of engineering, sometimes with terrible consequences, as the dams later gave way and caused the deaths of hundreds of thousands of people.

The first People's Commune was officially established in Gaoping township in Wushan county on September 10, 1958. By mid-October the last of the agricultural collectives in the county had been combined into communes of thousands of families, often made up of what had formerly been rural districts. For a few years, members were expected to eat all their meals in much-hated communal canteens, leaving them more time to work undisturbed by household chores. Wushan alone had over 3,900 communal eating facilities. As part of the disastrous policy to industrialize at all costs, approximately 100 million peasants were taken away from agricultural production by the end of the year to make iron. Unattended fields withered, and much of the 1958 fall harvest rotted in the fields. To keep the furnaces burning, peasants cut down trees, stripping the forests and causing the erosion of mountainsides that would later contribute to severe flooding along the Yangtze. Vegetable gardens were outlawed. Pragmatic Sichuan peasants began to sing a folk song updated with the words, "It's better to make a small dumpling than to work for work points." The small dumpling was a baby who would have a full grain ration from birth, a huge boost to the diet of the rest of the family in regions where grain was still available. The pressures of unrealistic government agricultural targets resulted in exaggerated and fabricated production figures and reports of bumper harvests that did not exist. Officials who questioned the policies or reports of unbelievable crop and steel production were dismissed, and peasants were beaten for not handing over what they did not have. In winter famine struck.

Mr. Li, now the owner of an excellent Sichuanese restaurant in New York City, was a child living on the outskirts of Wanxian in the late 1950s. He remembers looking with classmates for pieces of tin or old nails and bringing them to one of the neighborhood furnaces that burned through the night, melting down scrap metal found by people living on the street. His father, a mid-level pump repairman, earned enough for a comfortable life with all necessities, if no luxuries. Li was just starting school when the food began to disappear. This sudden absence of meat and vegetables in the shops and marketplace made such an impression on him that to this day he still remembers every item his mother bought and how much it cost.

Looking back, he says that though food became scarcer, conditions never seemed that bad in Wanxian city. "You were always hungry," he said, but he never heard of anyone starving. It was more that the food he was accustomed to was no longer there. At first, in the early winter of 1958, the family ate sweet potatoes instead of rice, and then, as these became harder to get, he and his mother started going into the woods to gather roots and fungus off the trees and to look for a kind of plant called bird vine that she boiled into a stew. What was frustrating, even as a child, was to know that good things to eat were there if you had money. It was just a question of how much and who could afford it. Eggs cost 2.5 yuan each, more than 10 percent of a young worker's salary of 18.5 yuan a month, but they were something Li's family could afford once his father was promoted to a more senior level with a salary of 55 yuan a month. Housing was practically free, clothing costs were minimal, and with few goods available for sale, there really was almost nothing to spend money on anyway, except food. For the majority of the ordinary mid-level workers, obtaining any kind of meat was almost impossible in 1959 and 1960. High-level cadres or people with well-connected friends had meat when there was no meat, grain when there was no grain, and that's just the way it was. Ordinary people had no idea how long the shortages would last, so families hoarded anything that would keep, further reducing what was available.

In the countryside, where Mr. Li often visited relatives, the situation was different and terrible. Children were swollen with malnutrition and peasants stayed alive only by collecting tree bark and roots. For them, Li realized, this was not a game, as it had been for him when he and his mother went off to search the fields and forest for food. In the fall of 1959, when the rice dried up and died from drought before the harvest, families took the rice husks, which were usually fed to the pigs, and ground them into a powder to eat. This gave everyone a stomach ache. The pigs grew thinner and were soon turned into the last bits of meat most people saw for the next two years. Some families ate leaves, others made do with the pulp left over from making tung oil, and many were constantly sick from eating plants and plant materials that were not digestible and often dangerous. Li remembers that everyone in the surrounding countryside looked exhausted and sick.

Despite the famine and the political movements of his early childhood, Li's family survived the years in good condition. Neither intellectuals nor outspoken, his family was never targeted or bothered in the 1950s. For him, childhood along the Yangtze meant playing along the river's edge and sliding down the rushing water of the city drainage ditches into the murky waters. All the boys looked forward to crab egg season. During the Great Leap Forward, these too were scarce and cost as much as four yuan a catty, which seemed like a huge sum for something you found in the river. This was a delicacy everyone longed for, but no family could afford to buy them in the market. After a big rain, the boys would scoop up the crab eggs from the shallow edges of the river with a net, and then catch the crabs with their hands, grabbing them around their middles to avoid their pincers. Li was frightened of being bitten but they were so good it was worth the risk.

Li also recalled the pleasure of taking time away from lessons at school to go out into the open spaces of the town with teachers and classmates to beat on plates and shout at flocks of sparrows attempting to alight. The children would rush from place to place, frightening the birds away each time they tried to land, until the birds eventually dropped dead of exhaustion. This exciting, but nonetheless required, activity was a part of the campaign to exterminate the "four pests" (sparrows, rats, mosquitoes, and flies), and every family had a quota of dead bugs and rodents.

In the evenings, the air was filled with dragonflies. There were at least three or four kinds, each of which had to be trapped in a different way. On the outskirts of the city, beekeepers tended their hives in fields on the edge of the forest and the boys tagged along after them. When the men were not looking, Li and his friends poked sticks in the hives and carefully stole the larvae. It took months of practice to do this without getting stung. Then they would catch crickets and grasshoppers and take them home. Running through the dried-out fields and climbing into the nearby hills, the young boys kept an eye out for poisonous snakes while looking for the *shejing*, the crown snake, that old men in the villages had told them about, a huge and monstrous snake with a crown on his head who had lived in a deep cavern for thousands of years. No one was sure what he really was or did, except that he had been there for all time, and was perhaps not harmful though he would be fearful to come upon.

During what became known as the "three hard years," from 1959 to 1961, estimates of the number of people who starved to death in the P.R.C. range from 20 to 45 million. An estimated 7 to 9 million people, out of a population of 70 million, are thought to have died in Sichuan alone. The county histories from the region describe the worst time, in the fall of 1959, with a few terse sentences. Wushan's gazetteer says, "Throughout the county, the majority of commune canteens stopped serving food. People had swollen bellies and began to die. In Guangyang and Dachang, hungry wolves often attacked the people and livestock. Fearing that they would be attacked as rightists, the cadres said nothing until this time."[2] In November a debate began about whether to accept relief supplies from the state. With the decision that the communes were no longer self-sufficient, grain shipments began from Chongqing to towns downriver. Several met with disaster when they foundered on rocks, and in one case a cargo ship with over a million catties of grain was lost.

The years following the Great Leap Forward were relatively quiet, with a welcome dearth of political campaigns. Peasants went back to farming in a normal way, and though the commune system continued, everyone went back to eating at home and no one had to spend their afternoons melting down their teakettles. This period of stability was not to last though, and only a few years later a new political campaign was underway. The Cultural Revolution, the result of a complicated power struggle between Mao Zedong and other party leaders, began in the spring of 1966 and by summer was in full swing.

From the county records one gets a limited but important glimpse of what happened, what was considered important, and what historians have dared to write. The official gazetteers provide an outline of the ten years between 1966 and 1976 but offer neither context nor commentary. On June 4, 1966, in Zhongxian, Cultural Revolution activities were initiated at all local teacher training schools and high schools to "carry out the work of the Communist Party." Schools closed soon after this and did not open again for almost three years. In Wushan, the records state that in early June "the Cultural Revolution began in Wushan and the Red Guards went all over, destroyed temples and shrines, burned old books, beat up leaders and cadres in the name of the 'revolutionary movement.'"[3] In July, the Oppose the Four Olds (old thought, old customs, old culture, and old habits) campaign was launched in the gorges, and rampaging students began a campaign of destruction.

They soon had company. Starting in the early fall of 1966, the state-run shipping line, newly renamed the East Is Red, brought hundreds of thousands of young people to the interior from all over China. Set loose by the early Cultural Revolution policy of free transportation to bring revolution everywhere, secondary students did exactly that, sightseeing and occasionally contributing to the general chaos wherever the passenger liners stopped. The newly formed Red Guards, made up initially of radical leftist students with revolutionary family backgrounds, had received Mao's encouragement to go out and destroy the Four Olds. As noted by the Wushan archives, the implementation of this policy took many forms—ransacking people's houses for classical or foreign books and other signs of less than revolutionary thought, wrecking temples, churches, and artwork, and cutting off women's long hair.

"Feudal" place-names were changed, and it was during this period that most cities acquired a main street called *Jiefang Lu* or Liberation Street. In Wushan, the twelve streets of the town named after the fairies who saved the area from flood and helped to create the Three Gorges were changed to names such as East Wind, Workers and Peasants, Red Flag, Oppose Imperialism, Grasp the New, Oppose Revisionism, and of course, Liberation. The fairy names were restored in 1982. Each town, however small, had its contingent of Red Guards by the end of the summer, and all strove to fulfill Mao's mandate. The Buddhas on top of the pagoda in Shibao Block were knocked to bits and the last monk who lived there was driven out. The pagoda itself was put on Zhou Enlai's list of protected cultural locations, which saved it from destruction. The Tianchi Catholic church in Zhongxian county, built by French priests in 1901, was left alone because it had been turned into an old-age home a few years earlier, but the grounds and outbuildings were pulled apart.

Though the violence in the other river cities never reached the levels that it did in Chongqing, where warring factions fought against one another with heavy weapons stolen from armament factories, each new political crusade increased factional fervor and demanded new targets.

In a region where roads were almost nonexistent and navigation still dangerous, the tremendous movement of people was in itself amazing. The highlights of 1967 and 1968, the two worst years of the Cultural Revolution in the Three Gorges area, give some sense of the destruction and unleashed energy of the time, as dozens of Red Guard groups tried to outdo one another with their political frenzy and loyalty to Mao. In 1967, in Wanxian county alone, the local government distributed 1.8 million volumes of Chairman Mao's *Quotations* and over 2.8 million pictures of him, at a time when the county population was about 800,000.

On August 1, 1967, with numerous Red Guard factions run mostly by senior middle-school students still springing up throughout the city, the Wushan County Military Bureau announced that it would support the Red Flag Revolutionary Clique as the leading group in the county. Over the next few weeks bureau staff and students joined forces to establish an armed militia base that became the headquarters of the newly founded Use Words to Attack and Arms to Defend Unit. This unit was soon busy throughout the gorges. On September 13, Red Guards from various armed groups went to the towns of Dachang (one of few intact Ming dynasty communities in Sichuan, on the shores of the Daning River) and Damiao (to the southwest) to attack and struggle against "evil leaders" and "capitalist roaders." "Struggle sessions" in which individuals were accused of being rightists, capitalist roaders, counterrevolutionaries, Kuomintang agents, or any of a number of other things became a part of daily life. People were held for hours on end on newly built stages in gymnasiums and schools while hundreds of spectators denounced them and demanded confessions. The struggle sessions in Dachang ended in a violent free-for-all, with twenty-one people injured, twelve seriously, and seven permanently crippled.

In February of 1968, "encouraged by the masses," eighteen members of the Wushan Revolutionary Clique sailed to Fengjie to attack the Number Two Coal Factory's Red Company faction. This resulted in another violent conflict. In March, over a thousand members of the newly established Red Peasant Central Group belonging to the Changjiang Commune in Wushan county came into Wushan city and were attacked by an opposing group. This battle left three people permanently crippled. In April, citizens of Yunyang county, upriver from Wushan and Fengjie, organized the Red Cloud Combat Group.

Joining forces with the Wushan Central Red Flag Group, the leadership group of the larger Red Flag Revolutionary Clique, they broke into the Wushan County Military Bureau's headquarters and stole a large amount of light artillery, including twenty-nine machine guns, fifty-seven rifles, forty-two handguns, and thousands of bullets.

During the summer, about 450 members of Wushan's Use Words to Attack and Arms to Defend Unit organized three armed attacks on Yunyang. In August, Red Guards from the Yunyang Chairman Mao Propaganda Team, allied with the Wushan Central Red Flag Group, came to Wushan, and the two groups and other similarly-minded bands of workers and students destroyed all the signboards of the Peasants and Workers Central Group and trashed their offices. They then beat up many of the county's so-called evil leaders and capitalist roaders who were already imprisoned in county kindergartens, primary schools, and high schools. Some one thousand people were injured. At about the same time, an armed group from Zhongxian attacked Fengjie and Yunyang, and another young militia group set off for Yunnan province, from which they did not return for a year.

In her memoir, *Wild Swans*, Jung Chang writes of her own trip along the river at the end of the Cultural Revolution.

> Wherever we went as we traveled down the Yangtze we saw the aftermath of the Cultural Revolution: temples smashed, statues toppled, and old towns wrecked. Little evidence remained of China's ancient civilization. But the loss went even deeper than this. Not only had China destroyed most of its beautiful things, it had lost its appreciation of them, and was unable to make new ones. Except for the much-scarred but still stunning landscape, China had become an ugly country.[4]

It is still a period that makes many people deeply uncomfortable and not one they much want to talk about. The short story "Wind and Rain in Pear Tree Town," written by Yuan Ran, is about the consequences of the visit of a high-ranking cadre to the town. This grim little tale highlights the destruction of people's lives and the loss of what was meaningful in them for the sake of illogical policies and empty slogans. The story refers to the ongoing campaign to "learn from Dazhai." The Dazhai Brigade in Shaanxi province became a nationwide model because of its members' commitment to collective farming and revolutionary principles, which supposedly made possible great achievements in agriculture. Most of its production figures proved to be false. In the spring of 1967, all production brigades in the Three Gorges region adopted the Dazhai method of assigning work points, that is, by an evaluation of the farmers' political attitudes rather than the amount of work accomplished. Cash crops were also forbidden. This had a notable impact on the peasants' enthusiasm for work, and the level of production dropped sharply.

> The people of Lishuwan lived from their pears. Pear trees were everywhere, and when the flowers blossomed, it was like a sea of flowers. That year, just at the time the trees bore fruit, a dozen cars came to the town. The Red Guards quickly went to tell the party secretary about this unexpected news. After a while, the townspeople gathered around together at an old pear tree, and after thinking over what to do, they decided that a young woman named Lixue should represent the people and offer the best and sweetest pear to this leader of the revolution who had arrived in a car.
>
> A short and fat middle-aged old man came out of the gorgeous car. He had a sharp look. The people of Lishuwan did not dare to look up at him. Everything was quiet and the people were afraid of him. Then Lixue came out. She offered him a plate with both her hands. A big pear was on the plate. Reflecting the sunshine, it was so beautiful. The middle-aged man laughed a little. He patted her shoulder. His expression was unattractive.
>
> Lixue had been born and brought up in the pear garden. She was pure and beautiful like the incarnation of pears. A uniformed man came and said mechanically, "Come with us. There are some problems that need to be explained clearly." The party secretary had lived a life full of trouble, so he knew what this meant. He wanted to stop what was going to happen, but when he tried to intervene, he was hit and knocked to the ground.
>
> Lixue did not come back. Instead, armed men came to the town and purged Lixue's relatives. They also ordered the townspeople not to work in the pear gardens because it was "capitalism." They ordered the people to work hard in the rice fields and "learn from the Dazhai spirit."
>
> The party secretary's nephew was working for the armed police. The secretary gave him too much to drink and learned the truth from him. The middle-aged man had forced Lixue to be a "model" and she resisted furiously. She tried to escape, but could not. So she hung herself.
>
> Black clouds hovered over Lishuwan. The town seemed deserted. The fruit ripened and just fell to the ground. The trees were not fertilized and died. People died.
>
> One night, the cry of a man was heard from the deep of the pear garden. The cry was like that of a monster. On the next day, the armed police personnel went into the garden and found the old secretary dead, embracing the root of a tree. The pear garden had been his life-long work. He and the pear trees went away together.

> Along with the old secretary, Lixue's seven-year-old daughter also died. There was nobody to feed her, so one night she went out to look for her mother and fell into a mountain chasm and died.[5]

By late 1968, the level of violence between competing factions had reached such a level that Mao sent the military into major cities to take control and restore order. The next step was to rein in the students by sending them to the countryside "to learn from the peasants." East Is Red ships again steamed up and down the Yangtze filled with young students being sent "up to the mountains and down to the villages." Tens of thousands of secondary school students from Chongqing, Yichang, and smaller cities were assigned to remote areas of Sichuan and Hubei provinces, many to the villages in the mountainous counties just north or south of the Three Gorges. Students from river cities like Fengjie or Wushan or Badong were also sent off to live with the peasants in outlying areas of their own districts and interior counties. Wushan's county records note that on March 11, 1969, the County Revolutionary Committee decided that students who graduated from the third year of middle school (grade nine) would not be permitted to join the military, look for work, or stay in urban industries, but must immediately relocate in the countryside. Similar directives were issued all over China.

Along the river, the children of government officials, factory cadres, and other workers in Fengdu and Fengjie and other small urban centers did not find it easy to adapt to the countryside and a way of life they had seen only from a distance. Unlike true big-city students, they at least were aware of rural life and what it meant. They had gone to middle schools where there were always some students from the countryside and had visited villages as children. For most young people born into educated families in Chongqing or Shanghai, the countryside simply did not exist, except perhaps as part of their parents' wartime stories.

Xiao Ma, a woman now living in New York City, boarded an East is Red ship in Chongqing in the fall of 1969 when she was fourteen and sailed off to an unknown destination in Wanxian prefecture. Entire grades of different middle schools were sent away together, so she traveled with a large group of her classmates who were to be settled throughout the Wanxian countryside. Unlike earlier programs, in which educated youth voluntarily went to work in the remote villages or on state farms, this was a required and supposedly permanent relocation which was to result in the movement of fifteen million young people to rural areas. Before they left, a teacher told them they were going to a place with mountains and water, *you shan you shui*, as he put it, an expression describing the ideal geographic combinations depicted in Chinese paintings. He had failed to mention that the mountainsides were so steep and the soil so poor that almost nothing grew, or that you had to leap over deep crevices that ran across the footpaths in order to get from place to place, or that poisonous snakes liked to sun themselves on the same paths. On the ship to Wanxian, the students traveled fifth class, a dirty, open expanse in the hold where people slept on straw mats if they had them. Xiao Ma and her classmates spent their nights huddled together in a heap on the floor for warmth, emerging only occasionally from the darkness to go look at some of the famous scenery going by and then to return downstairs to cry some more.

In Wanxian, a silent old man was waiting for them with a cart attached to the back of a truck. Lurching along in the open, they drove into the mountains on a curving dirt road with a long drop on the side. At each village, the inhabitants came out to gawk at this strange cargo bouncing by. At what seemed to be the top of the mountain, where the road ended, the driver told them to get out. He said nothing more and drove away. A few minutes later a group of ragged peasants swarmed around the bewildered students, grabbed their luggage, shouted something no one understood, and headed off along one of the mountain paths. Thinking their belongings were being stolen, the students ran after them. Only after one of the boys caught up with the villagers did they realize that these were the families with whom they were to spend the next year.

They reached the commune headquarters at dusk and spent the night freezing in the unheated administration building. The next morning, the students were divided into small groups headed for different brigades, made up of clusters of small villages. Despite the new political vocabulary and structure, old customs carried on, and the leaders of the village to which Xiao Ma was headed, though it was now officially known as a production team, came to collect and welcome her with great fanfare. The villagers walked, but they brought a cart for her with a marching band trailing behind blowing on horns and banging cymbals as if she were a bride in a traditional rural wedding. She had no idea why they were making such a racket. Exhausted by the trip, she asked them to stop, but the musicians told her they would not be paid if they were not loud enough. A specially prepared dinner of potatoes and pork fat was waiting. She thought it was disgusting.

The village had about twenty families, most of the same surname. Xiao Ma was the only student here, but others were close by in neighboring villages and they spent most of their time together. She was assigned to live in the home of a peasant family and started out by helping out around the house and in the fields, except that she wasn't much help as she had never even cooked dinner in her life before. The first useful thing she can remember doing was when an elderly neighbor died shortly after her arrival. Local custom decreed that one's closest relatives should be prepared for death by someone who was from far away, and so Xiao Ma was brought over to dress the dead body for burial. She was terrified at first, but it wasn't as bad as she had expected, and the family was always kind to her after that, even though she wasn't much good at growing anything.

Until she got off that cart in the middle of nowhere in 1969, Xiao Ma had no idea that anyone anywhere could be so poor, a reaction typical of many students "sent down" to the countryside who had never before seen rural China. Few villagers had a change of clothes and they stayed in bed if they washed their trousers, something that could not happen often. Rice would not grow here, so the farmers subsisted mostly on potatoes and millet gruel. No one had a pig. Most peasants kept a chicken or two and were permitted to sell eggs to the state market to earn a bit of cash, but this required walking several hours over the mountain to the commune headquarters. Once the students arrived with their small government stipends, the peasants sold their eggs to them and eliminated the trek. The students, who were hungry most of the time, had egg eating contests until they were sick.

Peasants in the Three Gorges regions remember with mystification the period when the students arrived. In Xiao Ma's village, peasants attended endless study sessions to hear about Mao Zedong Thought, but connections with the outside were weak. A few men had been to Wanxian, and one had inadvertently ended up in Chongqing because he had made the mistake of going to buy something in Wanxian in the mid-1940s and had been conscripted by the KMT. The peasants knew that the students had been sent to learn from them, but for what purpose, they had no idea. Chairman Mao, who knew everything, had decided this, so there must be some reason. When students from the cities first arrived, they asked the villagers why they did not more enthusiastically support the revolution. The peasants said they would struggle energetically once they no longer needed their energy just to produce enough to eat. Here, the hardship of life dimmed political zeal, something the students soon came to understand.

The policy of mandatory relocation for secondary school students had just begun the previous winter, and this was the second set of city teenagers the brigade had encountered. A group of senior middle-school students had been sent here a year earlier and had found conditions so unbearable that some of their parents with connections had managed to get them transferred out. The students had been difficult, but despite this, the peasants initially welcomed this younger group of students. The villagers' attitude changed as it became clear that few of the new arrivals were interested in doing anything more than their predecessors, and that their presence meant more mouths to feed, not more help. The girls were incompetent at household work or tending a garden, unable even to recognize the difference between a vegetable and a weed. The boys were even worse. Many refused to do anything at all and spent their days loitering about, breaking into sheds and empty houses. They stole potatoes and small items of no use to them but valuable to the peasants. Eventually the situation got so bad that the public security bureau and the local branch of the PLA were called in to identify the culprits and frighten them into behaving.

Over time, the relationship between the frustrated peasants and the unhappy students evened out somewhat, and both sides shared what they could. Antivenom serum was not available locally, but students from the city were provided with a supply that they shared with the villagers, who spent more time in the fields and were more often bitten by snakes. With nothing to offer in return, recovered farmers more than once resorted to killing their watch dogs in order to send a leg of meat to show their gratitude. Medicine, aside from the occasional boiled herb, was almost nonexistent, and there was no doctor for miles, so the students shared their supplies of antibiotics and iodine as well, hoping they were doing more good than harm. Many illnesses were treated by female witches called *wupo,* or men known as *daoshi,* who served as curers and exorcists. Called in when a ghost in the house was causing illness or harm to the family, the exorcist usually began by taking a few sips of wine, letting the ghost enter his or her own body, and then eventually spitting it out.

In peasant homes in remote villages in Wanxian prefecture today, a photograph of Chairman Mao hangs on the main wall of many houses. Across from the door, along with photographs of deceased family members, it is in the spot for the most revered of the departed. Old Li, a peasant from near Shibao Block, says Old Mao was a great man but a little fuzzy in his old age. Pressed for details, Li remembers the 1940s and 1950s with nostalgia, when there was something to fight for and life was going to get better. About the Cultural Revolution, he only shakes his head and grunts.

History here is closely guarded. Peasants, however, accustomed to saying what they like, tend to speak directly about subjects that officials might rather skirt. Preventing such openness provides local authorities with something to do whenever anyone from the outside is around. Then there are the official archives, to which only certain officials and scholars are granted access, usually after a lengthy application process and payment of hefty fees, and the published gazetteers, a noncontroversial collection of facts going back to the beginning of time, without commentary or analysis, bound in thick volumes available in Chinese libraries and bookstores. Many of the gazetteers are also available, as I soon discovered, though no one in the Three Gorges believed it, in libraries in Beijing, Hong Kong, and the United States.

Over the decades, most educated people in China have absorbed a continuing awareness of the subjective and complicated morass of what information can be shared and what not, and with whom. Information that you read in a newspaper or book can become a state secret if told to the wrong person, and transmitting even what seems to be public knowledge can cause immense problems for an unlucky person, particularly where foreigners are concerned. There is a Chinese proverb, *fei wo zu lei, qi xin bu yi*, which means "those who are not the same as I am have a different kind of heart." That is, they are fundamentally different, and one should always be careful of what is different, for it is impossible to understand their motivation or what they will do. This expectation applies in almost every dealing with outsiders, whether they are from another country or over the next mountain, and it added one more barrier to my attempts to obtain published materials and to go to locations long since declared open to foreigners in the Three Gorges.

One of the first times I was planning to spend a few days in Wushan, I met briefly with one of the young officials to whom I had been sent an introduction. He offered his assistance and said he would get me a copy of the local gazetteer. A few weeks later, when I returned, he suggested that as there was really nothing to see in Wushan that I hadn't looked at already, perhaps I should just get back on my ship and be on my way. When I pointed out that this was not possible (I was traveling on a tourist ship with a fixed day of return), he unhappily said in that case he would consult with the senior person, who happened to be his father, and took me to his office. In the meantime, yes, he did have a copy of the county history, but he didn't know if I could look at it.

We sat at his double-sided desk, he on one side, I on the other, with the big green county gazetteer between us. The electric fan clunked overhead. It was a hundred degrees and in the adjacent office young women were slumped over asleep at their desks for their midday naps. After a few phone calls, he told me that I could read the volume after all, but only in the office while he was there, as if his personal supervision would ensure that I took in nothing I should not. I picked up this 700-page book spanning 5,000 years of events listed in tedious detail and began, estimating that at the rate I read Chinese I would have to remain in his office for at least the next two years, something neither of us was likely to enjoy.

After a while, the elder Mr. Li showed up. He was not so bothered by my presence, no doubt because he had already dealt with every trouble a foreigner could bring and I did not rate particularly high on the problem scale. While no one in Wushan minded my visit, he explained, one always had to check at a higher level as well, and the higher level was not so thrilled. He said I could borrow the book and read it in my hotel room, or, no doubt knowing full well how unlikely this was, that I could go to the Xinhua bookstore and get one myself. Plus, later on in the week, I would be allowed to travel upriver on the Daning to the city of Dachang, as I had planned.

Old Mr. Li pointed out that I had a tourist visa, and on a tourist visa one could ask questions and write about sites of interest to tourists, but tourism no longer included history. Back in the early 1980s, when I first came to China and there was not as much to show people, it did. At that time, it was important to show foreigners the new China and how it was different from the old China. Now that the Three Gorges had modernized, it was no longer necessary to visit factories or to talk about the past. Now there were good restaurants and high-speed boats beyond the Little Three Gorges and an illuminated cave open to foreigners. I would see all these things myself during my trip, and to make sure it went well, I would not go alone. They would provide a man to protect me and a girl to talk to me. He concluded definitively, "It's not just you. Other foreigners do not always take local rules into account." Referring to an American colleague of mine, whose Chinese name I fortunately knew or I would have had no idea what he was talking about, he elaborated: "For example, your old friend, Hai Men Xiansheng, was here recently, and gave us a very big headache. He went where he was not supposed to go and we had to catch him." Guilt by association is hard to escape in Wushan, and the authorities were not ready for a second headache so soon.

The next day I went to the Xinhua bookstore and asked for the gazetteer. "That's a government document," said the clerk dismissively and waved her hands toward the top of the street, "You can only get that in the government bookstore. Everyone knows that." I got as far as the interior door of the government compound, where the bookstore was located, before I was stopped. I explained my mission and was shooed out the door by a young man who exclaimed repeatedly, "Foreigners can't buy books here. How could a foreigner read a Chinese book?" He and a half dozen other people stopped, frozen in motion, until I was gone. Back in my hotel room, I had an idea. Wushan now had privately owned office shops, where one could make long-distance phone calls and use the photocopy machines. If I could not buy the book, I could xerox it. I explained to the friendly woman in the family-run store down the block that I needed the entire book copied in ten days. Fine, she said, come back every day to check on it. On day one it was going well, but by day three it had become a lot to do, and she asked if I couldn't buy the book somewhere in town. I told her, Chinese style, that it was *bu hao mai*, not so easy to buy. On day four, the proprietress said she had heard it was available in the government store. On the fifth day, she offered to get it for me, and I gratefully accepted.

At the designated time a few days later, I returned to the shop. A young man whom I had not met before was there. He said his elder sister could not come in that day; it was inconvenient, but she had left me a package. I began to wonder if my circumventing the rules had brought her trouble. I paid for my book, which was handed to me in brown paper like a porno magazine, and I left her a glossy calendar with pictures of New England cows and mountains. The next day, I saw the shop owner and her brother through the window. He stuck up his thumb, signifying something good, and I took it to mean that no one in the family was being persecuted for their role in my recent purchase of the county gazetteer. She nodded and smiled, but looked away and did not rush to greet me as she had earlier in the week. I appreciated her help and I hoped she liked her calendar, but I knew it was best to take my business to another shop for a while. I returned the book to Mr. Li, explaining that I did not need it as I now had my own copy. He beamed and said that was good. He liked history too.

CHAPTER 8

DECISIONS AND DELAYS

THE HISTORY OF THE THREE GORGES DAM, a dam that is so altering lives along the Yangtze, is little known by the people most affected by it. In 1919, when Sun Yat-sen first proposed a dam that would one day change China, few people in the gorges heard about it or considered it anything but a distant dream. Now almost every peasant can tell you that it was Sun's idea, and that Mao Zedong himself would have seen the dam started and finished decades ago had other things not gotten in his way. Almost no one is aware of how much effort Chiang Kai-shek put into finding a way to build the dam or that the United States was deeply involved in this project in the 1940s. In the late 1950s, and during the later years of the Cultural Revolution, the question of whether and when to build the dam was frequently debated at the highest levels of the government. The repeated attempts to set a date to begin construction and the numerous times final approval was postponed reflected both the political turbulence of the era and Mao's persistent desire to dam the Yangtze.

In the mid-1950s Chairman Mao and other senior leaders repeatedly sailed through the gorges. The visits were never announced in advance, but news of the chairman's boat spread quickly and thousands of people turned out for a glimpse of the naval vessels, the *Yangtze*, the *Xijiang*, and other state ships. The importance of a place is in part indicated by the stature of its visitors, and noteworthy visitors were carefully tracked here. A locally published book, *The Dream Came True in Yichang*, lists the Chinese leaders who have inspected the gorges and potential dam sites over the past fifty years. From Mao Zedong and Zhou Enlai to Deng Xiaoping and Jiang Zemin, dozens of men whose names are household words in China stopped to look at Sandouping and Zhongbao Island and, later, to admire great chunks of sand and earth being dragged out from the riverbed with shovels and bulldozers.

The list is long and eclectic, and includes leaders whose fortunes veered sharply up and down after their Yangtze trips. Regardless of who they were, they said much the same thing about the project and what they saw. In 1960 Liu Shaoqi, then president of the People's Republic, who died in 1969 after years of attacks and imprisonment during the Cultural Revolution, stated, "The Three Gorges project is a great project, from which we can see a brilliant future for our country. The Three Gorges project is the center of flood control, hydropower, shipping, and water diversion, which will be of much benefit to the whole country."[1] After his death, a stone he had picked up on Zhongbao Island and kept for years was reportedly found among his belongings.[2] Twenty years later, Deng Xiaoping, another pragmatist, looked through a telescope at Zhongbao Island and the surrounding area. A day later in Wuhan he announced that the Three Gorges Dam would present few problems for shipping or ecology and would play a great role in flood control and energy generation. He added, "It is not good to rashly oppose the Three Gorges Dam."[3]

By the late 1950s, even though no progress on the Three Gorges Dam was evident, dam-building had nonetheless already impinged on the lives of families all over China. Between 1955 and 1960 thousands of men in the Three Gorges were drafted to work on dam-building brigades on the tributaries of the Yangtze and elsewhere. The stirring songs and patriotic verses of the time brought the future dam into the realm of general cultural knowledge and reminded people again of the beauty and power of the river, and of China as well. Following a tradition that goes back at least 1,500 years, many visiting officials wrote poems to commemorate their visits. Mao Zedong's poem "Swimming" was composed in Wuhan in 1956, at a time when the dam's rapid completion seemed an imminent possibility. The poem declared:

> A bridge will fly to span the north and south,
> Turning a deep chasm into a thoroughfare;
> Walls of stone will stand upstream to the west
> To hold back Wushan's cloud and rain
> Till a smooth lake rises in the deep gorges.[4]

Constantly quoted, it was a staple of the time, evoking a world irrevocably changed by the will of man.

Many of the works written in the 1950s mirror age-old poetic descriptions of the scenery and, as in ancient times, express contemporary political and social concerns symbolically. Poetry is an art that is far more part of daily life in China than in the West, and one not confined to the educated. Chen Yi (1901–1972), a native of Sichuan and member of the Communist Party since 1923, was the mayor of Shanghai from 1949 to 1958 and then China's foreign minister. He was persecuted during the Cultural Revolution and died in 1972. In "Visiting the Three Gorges Again," written in 1959 after an absence of thirty-two years, he described a voyage not only through the gorges, but through history to a time when it still seemed the dream of a Communist victory and all that it stood for might be achieved.

> I am glad to see again the mountains and river so magnificent,
> And I can still discern the bloodstains of the old dream.
> After passing safely thousands of dangerous rapids and shoals,
> The head water like a sword has pierced the Kuimen Gate.[5]

Throughout time, Chinese leaders hoped that a great dam would relieve the suffering and stop the destruction caused by the uncontrollable summer waters and provide the power for widespread economic development. Between the beginning of the Han dynasty in 206 BC and the end of the Qing dynasty in 1911, there were 214 major floods recorded along the Yangtze, averaging about one every ten years. Approximately once every century huge floods caused massive damage and loss of life all along the Yangtze, but the greatest destruction was usually in the middle reaches where the land is flat and heavily settled. Here large tributaries such as the Han and Huai rivers pour millions of tons of water into the Yangtze every summer. It is the location of these tributaries below the Three Gorges Dam and their role in flooding that have for many decades led some hydrologists to question whether any dam in the gorges can effectively deter the worst catastrophes that occur downstream. One of the most disastrous floods along the Yangtze was in 1788, when the peak flow in Yichang reached 3 million cubic feet per second (compared to a high-water norm of 1.4 to 2.5 million and a low-water flow of as little as 0.1 million cubic feet per second). The ancient Jingjiang dike, first built in AD 345 and repeatedly repaired and expanded, broke in twenty-two places, leaving the city of Jingzhou in the middle reaches of the Yangtze under water and causing great loss of life.

Two more catastrophic floods occurred in 1860 and 1870. During the flood of 1870, the worst since the tenth century, calculations made using remaining high-water marks indicate that the volume of the flow of water at Yichang was 3.9 million cubic feet per second. In Wushan and Wanxian, the water level reached about 150 feet above the normal summer high-water levels. The city walls, onto which town residents had climbed to escape the submerged streets, cracked and collapsed throughout the Three Gorges. The Yidou gate in Fengjie, built in 1474, gave way under the weight of the water. In Wanxian, over a hundred thousand people were left homeless and without means of support. The entire county of Fengdu was flooded with only the tops of the Buddhist temples visible on Mount Mingshan.

In the twentieth century, there was an almost record number of serious floods, due in part to the over-cultivation of the riverbanks as the population increased and later to the destruction of forests and misuse of land during the Great Leap Forward and the Cultural Revolution. The flood of 1931 affected an area of 8.4 million acres, an area approximately the size of New York state, and submerged 7.4 million acres of farmland, destroyed 1.8 million houses, and caused the deaths of 145,000 people. The city of Hankou was flooded for over three months, and then four years later, in 1935, 3.74 million acres of land were flooded again, and another 142,000 people perished.[6] Though almost all the great Yangtze floods affected the Three Gorges to some extent, the ones that caused the most damage in the middle reaches were not necessarily as serious upriver. Likewise, there could be terrible flooding in the narrow gorges yet relatively little harm done downstream once the river widened below Xiling Gorge. The 1954 flood, for example, one of the worst of the century, was no worse in the gorges than the 1956 flood, which was scarcely of note along the rest of the Yangtze.

Until the twentieth century, most efforts to limit the dangers of the rapids were focused on requests for supernatural intervention rather than hydrology or engineering. During the flood of 1788, by order of the Qianlong emperor, nine iron oxen were cast and placed in the Yangtze. As water submits to iron and earth, and oxen belong to the earth, the theory was that iron oxen would control the flood, but this did not work. One of the first attempts to use modern technology was in 1897, after a terrible avalanche in Xinlongtan (New Dragon Rapids) in Xiling Gorge had caused a whirlpool that in less than a year wrecked over a hundred junks and drowned a thousand people. Local villagers claimed this was the work of an angered dragon who, starved of his diet of waterlogged corpses by recent improvements in navigation, caused the avalanche to increase his food supply. They believed he should be either placated or frightened, while officials saw a solution in removing the thirty-foot boulder endangering traffic in the channel. William Ferdinand Tyler, a young British deputy coast inspector and expert in engineering hydrology and explosives, was asked by the Chinese government to use dynamite to remove the rock without further angering the dragon. Eventually, after several attempts and the assistance of thousands of laborers, he succeeded.

Tyler later reported that one of the river dragons put in an appearance just as he was preparing the blast.

> I was working on the central rock and saw that there was great excitement on the groyne. When we met—the three of us—at lunch time the Scotsman said: "You won't believe what I am going to tell you, but I have seen the bally dragon." He had seen only its head—a monstrous snout of some six feet long with a tubular mouth like a fire engine suction pipe. Then my "boy" appeared and with great excitement told how he had also seen the beast—and he had seen its tail. . . . I never saw the creature; I watched for hours in vain; but it appeared occasionally round about the whirlpool for a period of some weeks. The Chinese swore it was the body-eating dragon that had caused the rapid; and undoubtedly it must have been a monster fish.[7]

Tyler estimated the creature to be thirty feet in length and concluded that the dragon was most likely a sturgeon, an enormous fish plentiful in the Yangtze at this time. Fishermen in Xiling Gorge commonly caught them about six feet long, while they were still young and tender. The fish, often available in the marketplace in Yichang, was a local specialty.

In the 1920s and 1930s, with blasting rarely an option because of the technical difficulties and frequent military skirmishes, authorities instead focused on other methods to improve navigation. Between 1928 and 1937, nineteen winching stations that used ropes and pulleys to pull large steamships through the rapids were set up between Yichang and Yibin. A version of an old junkmen's song from Xiling Gorge asked heaven to protect them as they sailed through the Xintan rapids, "for if the dragon king gets us, then boat and boatmen are finished." The words of a 1950s version, rarely heard in recent decades, were modified to reflect a more modern era. "Cables and winches help us pass through the shoal and our machines rattle the palace of the Dragon King. We glide into Sichuan even before the all-clear siren sounds."[8]

In 1949, there were major floods in the Three Gorges. According to Cao Yinwang, in *Zhou Enlai and Water Conservancy*, Mao wept when he heard that "many villages were devoured in the water and surviving villagers were bitten to death by snakes."[9] The new Communist government immediately focused on flood control, water conservancy, and ways to improve the safety of navigation. Over the next ten years, approximately fifty-six large reservoirs, 4,000 large irrigation projects, and 19,000 small- and medium-size water projects were completed.[10] In the Three Gorges, the first government engineering team arrived during the winter low-water season of 1950 and began blasting out the reefs and shoals that had made travel so hazardous for thousands of years.

In Xiling Gorge, the well-known and greatly feared Come to Me Rock, in the Kongling rapids, was one of the first to go. A nearby section of the river was called the Gate of Hell because of the vast number of people who had drowned here. The rock took its name from the story of a boatman and his young son who had been ordered by a landlord to transport goods to Sichuan in such a short time that they had to travel day and night. The father died in the Kongling shoal, leaving his son alone in the world. The boy, Spring Tide, stayed on to guide ships through the dangerous waters and became known in time as the son of the god of ships. One day an official with a cargo of stolen gold and jewels tried to force the boy to steer his boat for him. At the most dangerous spot, Spring Tide leapt into the river and disappeared, and a huge rock rose from the water and smashed the greedy official's boat to bits. Though the boy was never seen again, ever after that, the sound of the water splashing over the rock cried out "Come to me." If vessels steered straight at it, they could pass by the shoal safely. If they attempted to pass it, the result was disaster.

One of the most famous catastrophes in Xiling Gorge occurred in 1900 when the *Suihsing*, the first German steamship to sail through the gorges, hit the rock and sank. According to a local version of the event, which rather closely reflects the legend, the Chinese pilot of the *Suihsing* steered straight at the Come to Me rock, knowing that this was how to avoid it. Not understanding this and fearing that the pilot was attempting to sabotage the ship, the German captain supposedly pushed him into the river, where he drowned. The captain himself then attempted to steer the boat away from the rock and as a result hit it, drowning himself and sinking the ship. In the Western version, the captain does not push the pilot out, which would have been difficult from the bridge of a steamship, but their fates were the same. The boulder ripped a huge hole in the side of the ship and, after drifting downriver out of control for almost half a mile, it plunged headfirst into the river. A number of people drowned, and more would have had it not been for the quick work of the lifesaving Red Boats, which rescued 285 Chinese and 33 foreign passengers.[11] To reward the quick work of the rescuers, the British consulate, the only Western consulate in Yichang, sent a donation of 300 silver dollars to the crews.

Soon after the Communist victory, Lin Yishan, the former party secretary of Liaoning province, was named as head of the Ministry of Water Conservancy and the Yangtze River Conservancy Commission. These were later reorganized into the Yangtze Valley Planning Commission, reporting directly to Premier Zhou Enlai. The first major project under his direction was the Jingjiang Flood Diversion Project. Work began in 1952, on the site of the Jingjiang dike, in the middle reaches of the river near Shashi and Jingzhou, in a winding stretch of the Yangtze locally called the Jing River. The flat Jianghan plain stretches out behind the dike, and when it is breached, the plain becomes a giant lake, drowning fields and farmers.

The diversion project was carried out by the "masses," thousands of untrained but at least officially enthusiastic "volunteers" mobilized for huge infrastructure projects. In the spirit of the time, 200,000 workers and 60,000 soldiers with buckets and shovels dug, in only seventy-five days, a reservoir capable of holding more than 177 billion cubic feet of water, enough to reduce the strain on the dike in the event of a major flood.[12] By comparison, it took workers, students, and cadres ten months to complete the Great Hall of the People in Beijing in 1959. Despite such projects as well as extensive dike repair and new construction, the hundred-year flood (i.e., of a magnitude that occurs once in a century) of 1954 left over 7.8 million acres of cultivated land along the Yangtze under water. The Jingjiang dikes held, vastly reducing the number of casualties, but over 30,000 people still died, and another 19 million lost their homes or businesses. During the smaller flood of 1931, 145,000 perished. In Hubei province alone, in 1954, 3.5 million acres of farmland were flooded and 5.8 million people lost their homes or property. The Beijing-Guangzhou railway was cut off for a hundred days. By current estimates, the 1954 flood would have caused at least U.S. $300 million in economic damage had agricultural and industrial development been at 1999 levels.

Rewi Alley, an Australian journalist who lived in China for decades, described, in the flat and ideologically correct terms of the time, one man's heroism in the 1954 catastrophe in his book *Man Against Flood.*

> Then this is the story of the Gorges in flood. Yang Kuo-ching was of poor peasant stock. He had suffered greatly in the old society though, by the age of twenty-one, he was already counted as a good junkmaster. After liberation, he had become master of a junk of some fifty tons, operating between Wanhsien and Shasi. He was always on the best of terms with his crew and rigidly observed all safety regulations. He also held a leading position in the navigator's section of the seaman's union, and was a model worker of Szechuan Province. He had just returned from Patung to Yunyang when he received a call from

> the local county cooperative to the effect that they had been asked to try to stop a big junk loaded with bamboo which had broken loose from its anchorage owing to the flood, somewhere above Wanhsien, and was now coming down the river out of control. It would certainly be wrecked in the rapids of the Gorges and its crew and cargo lost, if steps were not taken to save it.
>
> Yang Kuo-ching took fifteen experienced sailors and a cooperative worker out into the middle of the river in a small craft and soon spotted the big junk bearing down on them. They brought their boats alongside their quarry, and roped their craft to it. . . . On one occasion, they almost succeeded, but then a flush of flood water caught the junk and threw it out into the middle again. The salvage operation now began to take on more serious proportions. . . . Yang Kuo-ching . . . began to be worried about the safety of his men and finally decided the big craft was the safest place for them, as the small boats might turn over in the dangerous waters they were fast approaching. He, however, held on with all his might to the tow rope. He shouted to the others, "This is public property. We must not lose it!" But then came a terrific lurch as the big junk was caught by a wave, and Yang Kuo-ching was pulled into the flood, disappearing at once beneath the yellow current, which carried him below the surface far away from all possible help in a few brief seconds. The Yangtse is like that. Only his clothes which he had thrown off to allow for more freedom of working in the crisis, were left on the boat, a sad reminder to his comrades as they looked at them. The crew of both small and big boats redoubled their efforts in his spirit, however, and in the end all were saved through them. Next day, in Yunyang a memorial meeting was held by all local workers of a brave sailor and a true man.[13]

Two years before this flood, in 1952, while the Jingjiang reservoir project was underway, the Chinese government had revived plans for the Three Gorges Dam. The following year Mao Zedong took his first inspection tour of the Three Gorges. On February 19, 1953, Mao arrived in Wuhan by train and sailed westward on the *Yangtze*, accompanied by Lin Yishan, the director of the Yangtze Valley Planning Commission. During the trip, Mao told Lin that he had two priorities once the economy was stabilized—first to build a dam in the Three Gorges and then to build a canal system to divert water from the Yangtze to the dry north of China. It took decades longer than he had expected, but ground was broken for the Three Gorges Dam in 1994. In the year 2000, almost fifty years after he had suggested it, the government began work on the south-to-north diversion channel to bring water from the Yangtze to northern China.

Despite the potential expense and difficulties of the dam, the 1954 flood was a turning point in decision-making. On October 1, 1954, the fifth anniversary of the founding of the People's Republic of China, Mao made a personal request to Premier Nikita Khrushchev for Soviet technical assistance in building the Three Gorges Dam. Zhou Enlai followed up through diplomatic channels, and within months, the Soviet Union supplied a team of over a hundred people with a dozen airplanes to survey the Yangtze, adding to the topographic material already collected by the Chinese, Japanese, and Americans. An additional fifty-five experts joined the project in Wuhan over the next five years, while dozens of other engineers in the Soviet Union made preliminary engineering studies and designs for the dam. In 1960, shortly before the Sino-Soviet break, an American congressional delegation touring Leningrad reported that they saw Soviet engineers at work building models of the Three Gorges project.[14]

The political mood of the mid- and late 1950s was defined in China by its maniacal expression of power through rapid construction and the striving to establish supremacy over the environment. This disrupted not only the lives of the people forced to take part in these programs, but also damaged the mountains, rivers, and forests they set out to adapt for human use. China's leaders subjected the natural world to the revolutionary conviction that, in the same way neighbors could make steel from nails overnight at home, fields would produce ten times what they had the year before and riverbanks stay in place without the forests that had anchored them.

As a sign of its growing importance, the Yangtze Valley Planning Commission was in 1956 renamed the Yangtze Valley Planning Office, though still run by the same people, and charged formally with planning the design and construction of the Three Gorges Dam. Soviet and Chinese scientists attempted to come to an agreement on the ideal location and height of the dam. Most of the Soviet engineers, who had come from the Soviet Ministry of Electricity and had little experience in flood control, recommended a high dam of 654, 719, or 768 feet. Lin and the Chinese staff supported a 768-foot version. The location of the dam was a particular source of contention between the Chinese and Soviet engineers, with the Soviets advising construction in the narrow Yellow Cat Gorge section of Xiling Gorge. This location would avoid the flooding of any large cities within the Three Gorges but would inundate large areas of farmland near Yichang. Chinese engineers preferred Sandouping, the broadest point of the river within the gorges, though the Soviets feared this would reduce potential electrical output and worsen dangerous currents below Sandouping.

As a possible solution, Lin Yishan proposed the construction of a low dam at Gezhouba, forty kilometers downstream from Sandouping in order to provide additional energy and reduce potential problems caused by the increase in the water level from a high dam. After long negotiations, agreement was reached, and in an essay entitled "Deliberations over a Number of Questions Concerning the Yangtze River Basin Planning," Zhou Enlai reiterated that a 768-foot dam, built with Soviet assistance, would solve the flooding problems in Hubei and Hunan provinces, improve shipping conditions, and make possible the generation of 150 billion kilowatts of electrical energy, enough to electrify and revitalize all of central China. It would also be an important first step towards a south-to-north water transfer. Ready to dedicate all available resources to the realization of this goal, the government assigned seventy of the one hundred graduating students in China who had studied hydrology and engineering geology to the Three Gorges project.

Throughout his rule, Chairman Mao was taken with the grandeur of water-related projects such as bridges, dams, and canals and drew upon the symbolism of victory over water to draw attention to his own power. Some of Mao's ways of making himself seen and demonstrating his vigor did not always please his advisers. In the summer of 1956, he decided that he would swim in three of China's largest rivers—the Pearl in Guangdong province, the Xiang in Hunan province, and the Yangtze. Regional leaders feared that his drowning would become their responsibility, and no one was exceptionally happy at the idea of accompanying him into the murky waters of any of these rivers, which the most senior members of the government were obliged to do. He persisted, and in late spring, despite the globs of human waste and other muck around him, Mao paddled down the Pearl River. He followed this with a relatively uneventful swim in the Xiang, stopping to visit a village on an island where he had played as a child.

In June he set off for the Yangtze, combining a morning swim in Wuhan with a chance to meet with Lin Yishan and talk about the Three Gorges Dam. Mao, joined by Yang Shangkun, then the director of the CCP's General Office; Luo Ruiqing, the minister of public security; Li Zhisui, his doctor; forty bodyguards, and various other national, provincial, and local leaders swam or floated fifteen or so miles down the Yangtze, accompanied by a flotilla of small boats ready to pull anyone out of the river if necessary. "The Three Gorges will be gone," he reportedly said to Dr. Li, apparently without much regret, "and only a great reservoir will be left."[15] The following year Mao proposed a swim in the rapids of the gorges but was somehow dissuaded.

In January 1958, Mao called a meeting of provincial and national leaders in the city of Nanning in Guangxi province, at which he criticized senior officials, including Zhou Enlai and Politburo member Chen Yun, both of whom had urged caution in his plans for what would become the Great Leap Forward. At the same meeting Mao formally introduced the Three Gorges Dam to the Politburo for the first time. Mao saw the dam as a key part of the Great Leap Forward infrastructure development program that would allow China to surpass Britain in the next fifteen years, a goal that had been set the previous month. (Britain was chosen as a competitive target not only because it was an industrialized Western country, but because it had been responsible for the Opium Wars and the beginning of China's decline in 1842.) Bo Yibo, then the chairman of the State Economic Commission, was unenthusiastic about the dam yet did not dare to oppose it directly. He did mention to Mao that Li Rui, then the director of the General Bureau for Hydroelectric Construction in the Ministry of Electric Power, had his doubts about the project.

To the surprise of all, Mao summoned both Li Rui and Lin Yishan to Nanning to present their opposing viewpoints. Lin Yishan spoke first, spending two hours making his case that the Three Gorges Dam would prevent floods in the middle Yangtze, improve navigation and irrigation, and provide hydroelectric power to spur development. Li Rui, who was to spend the next forty plus years opposing the dam, spoke for half an hour on what he saw as the failings of the dam, explaining that he did not believe it could solve the flooding problems and that alternative methods, such as dams on tributaries, would make more sense.[16] Then, reminiscent of fairy tales in which the king sets forth a series of tasks to be completed within a fixed period of time, Mao told Lin and Li to come back in three days with written essays. On the third night, the two papers were distributed to the assembled Politburo members and reviewed by Chairman Mao. Mao reportedly criticized Lin Yishan's paper rather harshly, praised Li Rui's and appointed him as a secretary for industry, and then, despite some hesitation, continued to support Lin's plans for a large dam in the gorges.[17] Li Rui never wavered in his reservations about the Three Gorges Dam for the rest of the twentieth century.

A month later, from February 26 to March 5, 1958, Zhou Enlai led a delegation of over a hundred Chinese and Soviet experts and officials on a survey tour of the Three Gorges. While sailing upriver, Zhou held a meeting to discuss the proposed Danjiangkou Dam and reservoir in the mountains of Hubei province, on the Han River north of Wuhan. The assembled officials passed a resolution calling for construction of the Danjiangkou project as *lianbing* (practical training) for the Three Gorges Dam.[18] Charged by Mao with keeping up the momentum for the Three Gorges Dam, Zhou soon held another series of meetings to bring China's leaders to a consensus about the project. With the demand for immediate and monumental achievements intensifying, virtually everyone concerned, except Li Rui, professed approval for the dam.

Facing heavy political pressure, Li Rui too ultimately concurred that the project was good "on the whole" and should be built "when conditions are ripe."[19]

A few days later in Chongqing, Zhou and other senior officials completed a draft report stating that all relevant experts and officials agreed that it was possible to build the dam. He and Mao then headed to Chengdu for a continuation of the Nanning meeting. Here Mao again criticized senior leaders responsible for economic development and busied himself reading classical works on water conservancy and visiting nearby Dujiangyan, the still-functioning two thousand-year-old irrigation project. On March 25, Zhou presented the draft report about the dam and a resolution was passed stating that the dam should be no more than 654 feet high, the preliminary design work should be completed by 1963, and "all relevant work should be done in accordance with the principles established by Chairman Mao."[20] Then Mao took the train to Chongqing and set out on the river. A local book, *The Dream Came True in Yichang*, describes the trip.

> At the end of March, 1958, after the Chengdu Meeting held by the CCP Central Committee, Mao Zedong, along with Li Jingquan, the Secretary of the Party Committee of Sichuan Province [and others] inspected the Three Gorges by the ship Jingxia from Chongqing for the first time on March 29th. When Chairman Mao himself saw the dangers of the shipping impeded by Yanyudui, a giant reef, he said to Premier Zhou, who was on the same ship, "For thousands of years, Yanyu Rock has impeded the safety of shipping, why not bomb it?"
>
> And it was at the same time that Chairman Mao listened to many suggestions about constructing the Three Gorges Project put forward by experienced comrades, such as Mo Jiarui, Li Jicheng, the captains of the ship, Yang Dafu, the chief pilot, Shi Ruoyi, a woman, the third mate, and He Lifu, a young helmsman.
>
> Also, he listened carefully to many suggestions on migrants, generating electricity, flood control, etc. made by local officials such as Ren Baize, the Secretary of the Committee of Chongqing Municipality, Chen Junqing, the Secretary of the Committee of the Party Committee of Fuling Prefecture, Yan Hanmin, the Secretary of the Party Committee of Wanxian Prefecture. Then he said very seriously to them, "Let thousands of people migrate from their homeland, it is a great contribution made by a generation to our country."
>
> At dusk on March 31st, the ship Jingxia made its way toward Zhongbao Island. Mao Zedong, standing on the deck of his stern, looked at the little island carefully with the telescope, and then listened carefully to the experts on why constructing a dam would be better in Sandouping than in Nanjin Pass. And then he nodded and highly praised them.[21]

The Chengdu meeting resulted in an endorsement of the project from the very highest levels of the Communist Party and stimulated renewed preparatory efforts for the dam. Mao's trip also led to a number of immediate improvements in the safety of navigation along the river. Between 1950 and 1970, over one hundred hidden reefs and shoals were removed and a total of 141 million cubic feet of stone hauled away.[22] Per Mao's suggestion, the *Yanyu*, or Goose Tail Rock, a huge boulder at the mouth of Qutang Gorge, was blasted away. "At the western entrance standing boldly out in the stream like a fort guarding the gorge," wrote Cornell Plant, "is the great rock Yen-Yu-shih. It towers some sixty feet high above the water during the low level of winter, and according to legend, stands on an immense tripod of rock, beneath which is a famous dragon castle, Kwei-long-kung."[23] It took seven tons of dynamite to remove it.

While sailing through the gorges, Mao also called for the electrification of the signal system. During the Sino-Japanese war, Du Sixiang, known as one of the greatest Yangtze pilots of all time, was the first person ever to navigate a steamship through the Three Gorges at night. Successfully avoiding Japanese bombs, he delivered much-needed supplies to Chongqing. In 1950 navigation safety was improved with the installation of 7,000 kerosene markers all along the length of the river, making possible night travel, which had been too dangerous to attempt except in wartime. Although the kerosene markers were of great benefit, they were dangerous and required constant attention. Within a year of Mao's trip, electrical markers suitable for use in the gorges were developed. In the 1970s these were replaced with neon lights that turned on automatically as dusk fell.

Many of the signal stations in the gorges, located in areas where only one-way traffic is possible or where the fog is exceptionally thick, were also set up in the 1950s. In ninety spots between Chongqing and Yichang, station keepers kept in constant contact with one another and ships on the river by telephone or ship-to-shore radio to communicate water, fog, and traffic conditions. The stations were usually high in the cliffs in lonely and difficult to reach places. The Laoguanmiao Station in Qutang Gorge, for example, was 654 feet above the water and manned for decades by three workers who had to climb down the stone steps they had chiseled out of the cliff to get their water. The Wushan station was one of the most isolated, located high above Wu Gorge. An excerpt from an article published in the mid-1970s stressed how a young man's developing sense of responsibility overcame the potentially dangerous boredom of the job.

> Chingshihtung Station was built on a steep slope above Wu Gorge, a desolate spot with only eagles and wild goats for company. An old worker and a young one run this post. When the young man first came, he was restless and complained that his duties were too simple. One day he sighted a ship approaching from upstream. He was about to give the signal to pass when a raft suddenly entered the narrow mouth of the gorge. He panicked and forgot what to do. The veteran worker immediately gave the warning signal, then used the incident to point out the high responsibility and skill the job required.
>
> One day during a storm a rockslide broke their telephone line. Navigation service could not be suspended to wait for a repairman to come a long way in such weather. Taking pliers and wire, the young worker managed to reach the cliff in a small boat. Climbing up with great difficulty, he found the broken line. But when he picked up the two ends a sudden numbness shook his arm. Realizing that a station along the river was trying to make a call, he stood the pain and swiftly spliced the line.
>
> Centuries of feudal and reactionary governments never tamed the Yangtze Gorges. Socialist men have been able to bring them under control in two decades.[24]

The movie, *Clouds and Rain over Wushan*, made by the young filmmaker Zhang Ming about the same signal station in 1996, more than twenty years later, provides an interesting contrast. The young signal man in the film is also diligent but even more restless, led by boredom into endless card games and a complicated, pointless affair. The mostly forced enthusiasm of the 1970s has been replaced with the listlessness of the 1990s, as the signal man waits for the water to rise with the completion of the Three Gorges Dam and wonders what he is going to do with himself. With better communications possible, the number of signal stations was reduced to forty-five by the year 2000, but it is still an isolated and difficult existence for the men who work in them.

In the spring of 1958, two months after Mao's trip down the Yangtze, the Chinese Academy of Sciences and the State Science and Technology Commission formed the Science and Research Leading Group for the Three Gorges Project and submitted a detailed research plan for the dam. Mao and Zhou quickly approved this, and 10,000 engineers and experts, representing over 200 scientific and technical research institutions, were assembled to work on the project. In August, Zhou Enlai held another high-level meeting at the seaside resort of Beidaihe in Hebei province. Under pressure for greater speed, the group adopted a resolution to complete most design work on the dam within a year. Groundbreaking was tentatively scheduled for only one year later. In May 1959, 1,888 delegates from sixty-six governmental and academic organizations unanimously agreed that a 654-foot high dam should be built in Sandouping.

Internal political struggles in China were to derail plans for a quick groundbreaking. At the Lushan Conference in July 1959, a meeting of the Politburo and other senior leaders, tensions were severe between supporters of the Great Leap Forward and those aware of and willing to speak out about its increasingly tragic results. Li Rui, who had been appointed as a secretary to Mao eighteen months earlier despite his opposition to the Three Gorges Dam, now expressed his opposition to the Great Leap Forward. The result was that he lost his membership in the Communist Party and was sent to the countryside to do farm work. Two months later, in March 1960, the construction date for the Three Gorges Dam was postponed until 1961, but by the summer the extent of China's economic disaster was evident and even Mao admitted that it was not the time to start. Relations between China and the Soviet Union continued to deteriorate, and in June 1960 the Soviet Union abruptly canceled all aid programs in China and withdrew its experts. A month later, the Three Gorges Dam project was officially slowed down and within a few more months it had ground to a complete standstill.

By early fall of 1960 most of the Yangtze Valley Planning Office (YVPO) employees had disappeared and all but forty scientists had left in search of food. To keep the YVPO from closing down entirely, Zhou Enlai granted a loan of 500,000 yuan to the office to enable the staff to cultivate farmland so they would not starve, and thus kept some parts of the office intact. Construction of the Danjiangkou Dam, the practice dam, had begun two years earlier with great hopes that sheer willpower would overcome other difficulties.

> The project meant cutting the tremendous current of the Han River, moving thousands of cubic meters of earth and stone and pouring over three million cubic meters of concrete. This usually requires modern machines. But they were not all available and the builders would not wait. The Party had called upon the people of China to maintain their independence, keep the initiative in their own hands, rely on their own efforts and build up the country through hard work and thrift. . . . There were not enough pneumatic drills to make blast holes. They relied heavily on sledgehammers. Picks and shovels did the work of earth diggers. Carrying poles took the place of dump trucks.[25]

The Danjiangkou project was also suspended from 1960 to 1964. Finally completed in 1974, it required the relocation of 350,000 people.

Though the Great Leap Forward was over by the end of 1960, it left China's economy in shambles. Some research continued, but the YVPO did little for the next few years while the country stabilized and recovered. After a gradual revival of the Chinese economy, Lin Yishan suggested in March 1966 returning to work on the dam, but Mao apparently was not convinced that all the technical issues had been solved. Feeling threatened by political rivals, he had others things on his mind. In May 1966, the Cultural Revolution began. On July 16, Mao took another swim in the Yangtze, this time in Nanjing, near the "anti-imperialist, anti-revisionist bridge" then under construction. The Chinese media reported that he swam fifteen kilometers (nine miles) in an hour. *China Pictorial*, which covered the event in the October 1966 issue, featured pictures of the thousands of people swimming alongside him with banners to celebrate the event. One poor man "became so excited that he forgot he was in the water. Raising both hands, he shouted: 'Long live Chairman Mao! Long live Chairman Mao!' He leapt into the air but soon sank into the river again. He gulped several mouthfuls, but the water tasted especially sweet to him!"[26]

Work stopped again at the Yangtze Valley Planning Office. Though differing on the question of the construction of the Three Gorges Dam, Lin Yishan, the director of the YVPO, and Li Rui, both became targets of the Red Guards and were attacked and imprisoned along with countless others. Zhou Enlai issued a command to the Wuhan Garrison that the Yangtze Valley Planning Office must remain open, but he was unable to shield either Lin or Li. Imprisoned in Beijing in Qincheng Prison, used primarily for political dissidents, Li suffered little direct physical harm, but Lin was less fortunate. He was held for four years in the cellar of the headquarters of one of the revolutionary factions near the Yangtze, in a room into which the water constantly seeped, flooding the room and making the conditions horrible. "Are you not a water conservation expert?" asked his tormentors, "That is why we have invited you to stay in a water prison."[27]

Locked inside his watery cell, Lin persuaded a sympathetic young guard to get him some sand, cement, and brick fragments and used these to block the flow of water into the cellar—his own small dam. Frequently beaten and fearing that he might well be tortured to death, Lin kept himself alert and focused by mentally working on dam construction projects and jogging, believing that at some point Mao and Zhou would ask him back to continue his work on the Three Gorges Dam. Several years later this did happen, when Lin was seriously ill with cancer of the eye and Zhou arranged for him to be transferred to Beijing for treatment.

In 1970, while engineers at the YVPO were at political study meetings and attempting to avoid trouble, radical factions in Hubei province began to push for construction of a low dam, the Gezhouba in Yichang, as a way to gain favor with Mao. Zhang Tiyu, Mao's attendant for many years; Zeng Siyu, commander of the Wuhan Military Garrison of the Revolutionary Committee of Hubei province; and Qian Zhengying, a protégé of Zhou Enlai and at the time the highest-ranking civilian in the Ministry of Water Conservancy and Electric Power, led the way. As concerns about possible Soviet aggression were running high, Mao was uninterested in building an enormous and expensive dam in the Three Gorges that might well become a military target. He was, nonetheless, quickly won over to the idea of the Gezhouba Dam, which could serve simultaneously as the largest practice dam ever for the Three Gorges Dam and as a component of it. Although the Yangtze Valley Planning Office was by rights to oversee any new dam projects on the Yangtze, provincial leaders took advantage of the YVPO's weakened authority during the Cultural Revolution to make the argument that since the dam was entirely in Hubei province, it fell within the scope of provincial administrative control.

In the early fall of 1970, with the approval of Mao and Zhou, the Wuhan Military Garrison and the Revolutionary Committee of Hubei province submitted a technical report outlining plans for the construction of the Gezhouba Dam to the CCP Central Committee. Similar to the low dam that was proposed by Lin Yishan over a decade earlier to moderate the effects of a Soviet-designed high dam, it was to be 226 feet high, with a power plant producing a total installed capacity of 2.4 million kilowatts and an average annual electric capacity of twelve billion kilowatt hours. The initial proposal from the Wuhan Garrison stated that the dam, which would flood 1,416 acres of farmland and require the relocation of 13,000 farmers, would bring to fruition "the Great Leader Chairman Mao's great idea of building calm lake over tall chasms." If the dam were to be bombed and destroyed, only 10.6 to 12.3 billion cubic feet of water would flood the lower reaches, not enough to cause a "serious disaster."[28]

On December 26, 1970, on his seventy-seventh birthday, Mao gave his approval, with words that proved unfortunately prophetic. "I approve the construction of the Gezhouba Dam. What is proposed in the documents is one thing, unexpected problems and difficulties that will arise during the construction are another. Therefore, you must be ready to revise the design of the project."[29] Revise they did, again and again. The dam was built under the Cultural Revolution *sanbian* (three sides) engineering policy, referring to the practice of simultaneous exploration, design, and construction. Without any detailed blueprints, Zeng Siyu called for a "thousand people design," or "design by the masses," with preference given to the opinions of political enthusiasts rather than engineers.

On the night Mao made his decision to go ahead with the dam, over 10,000 people gathered at the Gezhouba headquarters in Yichang, bearing placards emblazoned with the words *zancheng xingjian ciba* (I agree to build this dam). At the groundbreaking on December 30, 1970, four days later, over 100,000 "volunteers" were at the site waving Chairman Mao's Little Red Book. Organized military style, more or less, by Zeng, they set to work. Within a few months, many of the more experienced hydrologists on the project had been sent away to farms in the countryside as a result of their bourgeois and rightist attitudes about how to build a dam. The result was that the design and construction of the dam was constantly modified as new information or problems were discovered. Liangwu Yin notes this in his "The Long Quest for Greatness." "The irrationality of revelatory zeal and huge-crowd strategy defied laws of science and technology. Soon, numerous problems in design, navigation, sedimentation geology . . . plagued the construction. Zhou Enlai, already a cancer patient, had to go personally to the construction site to investigate. By 1972, the whole project was in such a hopeless shape that the construction had to be suspended. Out of over one hundred thousand workers and soldiers, only thirty-five thousand were retained, 'awaiting orders' for almost two years while the redesign took place."[30] In November 1973, Zhou Enlai asked Lin Yishan, who had recently been released from detention in Wuhan, to become the chief commander of the project. Lin accepted, although he had opposed building the Gezhouba Dam when China already had the technology to go ahead with a larger dam on the Three Gorges. Construction on the Gezhouba Dam finally resumed in October 1974.

In late 1989, just over eighteen years after it began, the Gezhouba Dam was finished. Supposedly both Mao and Zhou were so fed up and disappointed by the problems in building it that neither ever mentioned it again after 1973. In 1976, the year in which they both died, Mao, perhaps expecting to live much longer, reportedly asked that the completion of the Three Gorges Dam be noted in his funeral eulogy,[31] but this was not to be.

CHAPTER 9

NEW CURRENTS

IN OCTOBER 1976, ONE MONTH AFTER Mao Zedong's death, the Gang of Four, the political clique that included Mao's wife, Jiang Qing, was arrested and blamed for much of the radicalism and violence of the Cultural Revolution. The fall of the Gang of Four ended the involvement of many of their followers, including Zeng Siyu, the military commander of the Hubei Revolutionary Committee, and many other senior officials on the Gezhouba Dam project. In 1978, the pragmatic Deng Xiaoping came to power and began a program of economic and social stabilization, permitting greater leeway for free-market trade and allowing peasant families to farm individual plots rather than work in the fields for points allocated by the brigade or commune, a move that eventually led to the dismantling of the commune structure.

Sailing westward from the mouth of the Yangtze near Shanghai in the mid-1980s, a few years after economic reform had begun in earnest, one left behind a city where change was everywhere, but no one yet knew how suddenly it was going to overtake the past. The last skyscraper had been built almost forty years ago, and everyone still wore blue. Five years later the city would be almost unrecognizable. Along the open expanses of the lower Yangtze the return to the household responsibility system was well underway. After decades of state-planned agricultural production with targets set for each commune, the family was once again the basic unit of production. On the tourist ships we anchored each day at a different stop. At each one, the old commune leaders explained that any family that had fulfilled its grain or vegetable quota could sell the excess in the newly opened free markets and keep the profits. In contrast to the existing state stores, with their limited and usually unappetizing selections of moldering fruits and vegetables at government-fixed prices, the new markets were filled with fresh produce at prices set by the people selling it. Incomes shot up for the first time in decades, and scores of newly built whitewashed two-story houses appeared along the banks of the Yangtze and in the villages. Families bought bicycles and carts, making transportation easier. Peasants soon branched out into repair work, tailoring, furniture making, and other small industries. In those days, Deng Xiaoping's new slogan was "To get rich is glorious!" and one of the first steps in that process was the acquisition of the "four things that go round," bicycles, electric fans, sewing machines, and washing machines. Life was becoming more comfortable.

As we traveled upriver, the sense of well-being was less noticeable. In a commune in Jiujiang, in Jiangxi province, we drank tea in peasant homes on a hill above the railroad tracks. The farmers first told us how much better life was since Liberation and then how many children had been killed or dismembered in the past year by the trains below. We stopped at the thermos bottle factory in Shashi, where workers walked through an inferno of flying shards and molten glass in cotton shoes and sneakers. Double Happiness thermos bottles and Mandarin Duck brand bed sheets, also made in Shashi, were traditional gifts for newlyweds. As symbols of eternal love and economic fortune, the demand for these had soared. In the darkness and heat of the factory, liquid glass bubbled in open furnaces. Walking backwards around the rubble, the men pulled the hardening glass into long tubes that would become the vacuums for bottles that kept water scalding hot for two days. The chalkboards outside listed the number of model workers, bonuses, eye injuries, and burns.

At the Gezhouba Dam site near Yichang, we stood around a model of the still unfinished dam with its flashing red lights indicating the completed sections of the project and what was yet to be done. Miniature boats sailed through tiny locks. A deputy head of propaganda first lectured us about the economic significance of the increased electrical output and then ended with the story of how, in the pouring rain, thousands of patriotic people had carried buckets full of rocks to close the final gap in the dam. Fifteen years and a world of change later, guides at the Three Gorges Dam would point at the twinkling lights of another scale model and proudly list Siemens, Caterpillar, and other foreign companies at work on the project. The thousands of people who had moved to Yichang in the 1970s and early 1980s to work on the Gezhouba Dam and accompanying improvements to the infrastructure had turned Yichang from a small town into a modern city.

Figure 9.1 Cargo houseboats going through the Gezhouba Dam

In the late 1970s, once economic reform was underway, the Chinese leadership began to stress the importance of technological breakthroughs made during work on the Gezhouba Dam and their importance for the Three Gorges project. Much as Mao had done thirty years earlier, Deng became an enthusiastic proponent of the dam, encouraging its construction to bring electricity and prosperity to China, and as a symbol of its modernization. In April 1979, almost twenty years after he had been appointed to head the Yangtze Valley Planning Office (YVPO), Lin Yishan once again submitted a report to the State Council advising a rapid start to work on the Three Gorges Dam. A few weeks later, at a conference convened by the State Council to select a dam site, Lin announced to the 200 people representing fifty-five institutions related to the dam project that the party Central Committee had decided to speed up the construction of the Three Gorges project. His words sounding wearily familiar to many of the delegates, Lin added that final design work would be completed in 1980 and construction would start immediately thereafter. As had been the case at similar meetings twenty years earlier, many officials believed the timetable to be too hasty to determine the full impact of the dam on the region, but did not dare say so directly. The conference ended two weeks later without any decisions.

Progress on plans for the dam ebbed and flowed through the early 1980s as the decision about whether to go ahead with the dam was shunted from agency to agency. In early 1983, the Three Gorges Project Preparation Leading Group was established to begin the organization of a special administrative zone surrounding the Three Gorges Dam. Later that year, the State Council approved plans for a 491-foot dam, and the Ministry of Water Conservancy and Electric Power was instructed to make immediate preparations for its construction by building roads, harbors, and power supply facilities near Sandouping. In 1984, with scientific evidence suggesting that a 491-foot dam would cause disastrous sedimentation problems for Chongqing, the city appealed to the government to reconsider the height and build a 572-foot dam, which would be more effective for flood prevention and power generation and cause less sedimentation.

The next year, in March 1985, at the Third Session of the Chinese People's Political Consultative Conference (CPPCC), an advisory group to the State Council, Sun Yueqi, the chair of the CPPCC economic commission, openly expressed his disapproval of the dam project. Sun, who was ninety-two at the time, had worked on the first plans for the dam under the KMT in 1932. At the conclusion of the meeting, he led the other members of the economic commission, all of whom were over seventy years old, on a thirty-eight-day tour of the Three Gorges. At the end of it, he appealed to the CPPCC Central Committee for a postponement of the project.[1] Opposition to the dam became more vocal within China at this time, particularly after a government-led forum to discuss the Three Gorges Dam was held in September 1985. Ostensibly convened to give people a chance to voice their opinions, government authorities were taken aback by the number of negative comments, as had happened so many times before in China. Public discussion among intellectuals and officials became more commonplace, and for the first time since the 1940s, the dam began to attract considerable international attention beyond the confines of hydrologists and dam builders.

China's efforts to gain international support and funding for the dam also brought foreign interest. In August 1984 the United States Bureau of Reclamation signed a three-year agreement with China's Ministry of Water Conservancy and Electric Power to provide technical assistance on the planning, design, and construction of the Three Gorges Dam. The following year, the U.S. Three Gorges Working Group, a consortium of the U.S. Bureau of Reclamation, the U.S. Army Corps of Engineers, two banks, and three private companies, conducted a technical review of the project, with an eye to its feasibility and the potential for American government and private sector participation. In 1986 China also commissioned a $14 million feasibility study conducted by a consortium of Canadian government and business organizations and funded by CIDA, the Canadian International Development Agency. Both the U.S. and Canadian studies initially supported going ahead with the dam, recommending the construction of a 607-foot version, but support from both governments was later withdrawn under pressure from environmental activists. After initially signing an agreement with China to assist in the construction of the dam, the U.S. Bureau of Reclamation terminated it, stating, "It is now generally known that large-scale water retention projects are not environmentally or economically feasible. We wouldn't support such a project in the United States now so it would be incongruous for us to support a project like this in another country."[2]

Disagreements over the Three Gorges Dam continued to grow in China as well. The decision to proceed with the dam led to a bitter struggle between the Sichuan and Hubei provincial governments over the organization of a Three Gorges Special Administrative Zone in the parts of Sichuan and Hubei provinces that would be affected by the dam. Hubei stood to gain significantly from investment and infrastructure development, while Sichuan would bear most of the brunt of flooding, sedimentation, and the problems of relocation. Each province wanted more control but less responsibility, particularly over moving and finding employment for the million and more people to be displaced by rising waters. Ever since the ancient kingdom of Shu evolved into the province of Sichuan, the two western-most gorges, Qutang and Wu, have been under its administration, while Xiling Gorge, to the east, belonged to Hubei. The Three Gorges have always been one geographic entity, but rarely a political one, a situation that has affected the overall development of the region, leaving Badong and Zigui counties cut off administratively and culturally from the Sichuan towns to the west and physically, by the hazards of Xiling Gorge, from Yichang and the rest of Hubei to the east. Tensions between the villagers living in the area on the border of what are now Sichuan and Hubei were often severe. In the poor areas on the outskirts of Badong and Wushan, farmers quarreled over land, firewood, and other limited resources. It was not until after the establishment of the P.R.C. that these problems were resolved with the intervention of the county governments.

Deng Xiaoping's solution to old and new tensions between Sichuan and Hubei was to create a new Three Gorges province, directly under the administrative control of the State Council. An area encompassing 32,506 square miles of the Three Gorges region as far east as Yichang and as far west as Zhongxian county was marked out in February 1985. Li Boning, vice minister of Water Conservancy and Electric Power, was authorized to begin the immediate organization of a provincial bureaucracy in Yichang. He was to become the party secretary of the province once it was formally established and was given the right to choose the governor. The first meeting of the Three Gorges Preparation Group was held in Yichang on April 15 to work out the details of administering an area that would have eighteen million people and responsibility for the largest population relocation in China's history. Soon thereafter, droves of senior officials went off to Yichang in hopes of being appointed to new high-level positions.

In the gorges, both the *xiao guan*, the cadres responsible on a local level for industries such as trade and tourism, and the ordinary townspeople were aware of this new administrative unit, but it made little difference to them. Aside from the occasional joking comment that even if they became "Three Gorgers," they would still be from Sichuan or Hubei, nobody paid much attention to the upcoming change. Occasionally someone would mention to foreign visitors that since the establishment of the new province had not yet been formalized, it was officially a state secret, though not one that they went to any pains to keep to themselves. Within a year, however, concerns about the dam emerged once more, accompanied by a growing frustration over what Sichuan and Hubei provinces would gain or lose. Deng's plan was abandoned and in

May 1986, the new Sanxia (Three Gorges) province was disbanded before it ever truly existed.[3] In one day, there was a mass exodus of officials as hundreds of cadres got on planes and flew back to wherever they had come from. Yichang was returned to Hubei, and Wanxian and its surrounding counties to Sichuan.

In the mid-1980s, one result of the attempt to establish the new province and to gain final approval for the dam was the increasing official attention to the extreme poverty that existed in the Three Gorges area at a time when other parts of China with better land and transportation were enjoying a surge in prosperity. Beyond the Gezhouba Dam, the terrain changes sharply as low hills turn to mountains, marking the beginning of a way of life remote from that of the cities or of the broad and plentiful fields of the lower Yangtze. Though the first experiments with the household responsibility system in China took place in Sichuan in 1978 and 1979 under the leadership of then provincial party secretary Zhao Ziyang, in the early 1980s the economic and political reforms that had already changed much of China had just begun to make their mark here. As elsewhere, the people who prospered in the Three Gorges were those who both had something to sell and lived close enough to a market to get it there. Farmers working on steep slopes inland even a few miles from the river might have the freedom to sell surplus crops and go into sideline industries but, with poor land and no transportation, it did them little good.

In downtown Wushan, there were a few state-run shops that sold farm equipment and grain, two desultory department stores sparsely filled with *riyongpin*, daily-use items such as thermos bottles and can openers, displayed in locked glass cabinets, and a few collectively run restaurants where dinner was served until six and the food was awful. Between Chongqing and Yichang, Wanxian was the only city where one sensed even a shadow of the energy that was becoming commonplace elsewhere in China. Here people were on the streets until all hours, and the market was filled with tangerines and local bamboo and rattan products and exotic animals for dinner. Though the Cultural Revolution had already begun to seem distant to visitors by the early 1980s, it was not to the people who lived there. The towns in the gorges were still emerging from economic and political disruptions imposed on a region that had never had a well-developed economy or educational system in place to begin with. Local people who had persecuted and exploited one another had to look at each other differently once more, as they had done after earlier political upheavals. Although the communities were small, the fights between different factions and the evening of scores did just as much harm here as in bigger cities and were no sooner forgotten. In time,

Figure 9.2 Zhongxian

however, everyone regrouped and went on with the future, but for many years a pervasive weariness lingered on.

The problems in the gorges resulted not only from geography, but also from a lack of systematic development or government investment in the region since 1949. For nearly half a century, from the early 1950s on, townspeople between Chongqing and Yichang lived with the remote but constant possibility of a dam. In 1953, a government directive was issued to city governments and other local organizations that no new factories or important structures were to be built below the anticipated rise in the water levels, which was then 623 feet. Although this regulation was only sporadically enforced, it still had an enormous impact.

Before the surge in economic activity in the 1990s associated with the construction of the Three Gorges Dam, there were three main periods of government investment and economic growth in Sichuan after 1949. The first was in the 1950s when the state invested heavily in the establishment of industrial bases in the Chengdu plain and other relatively accessible places in the province. Wanxian and Fuling, the most industrially developed cities between Chongqing and Yichang at the time, were excluded because of their location and the expense of providing adequate transportation and other necessary support for major projects. The second occurred in the mid- and late 1960s, as tensions with the U.S.S.R. increased, when the central government established what was called the Third Front in the interior of the country, where key industrial production centers were set up far away from the Soviet Union. Wanxian and other cities were again largely excluded from this influx of investment, this time because of the assumption that they were to be submerged. Some towns, such as Badong, which became a fairly important base of munitions production during this era, received some benefit despite plans for the dam. Chongqing also became a major producer of armaments during the 1960s and received significant central government investment in the development of heavy industry, including steel, machinery, and chemicals, but with the result that light industry and agriculture lagged behind.

The third phase of rapid economic development in Sichuan took place in the early 1980s when new policies led to immediate though limited improvements in welfare, but these were not sufficient to overcome the problems of geography and lack of education and training. The brief-lived plan for a Three Gorges province stimulated an interest in investment in new factories and other projects intended to employ migrants-to-be, but with its abandonment in 1986, and the timing of the dam uncertain, the region lapsed back into economic stagnation. This coincided with a general downturn in the Chinese economy, which had improved rapidly between 1978 and 1984, after which personal income, particularly in interior rural areas, either leveled off or dropped until the early 1990s.

In the mid- and late 1980s, the Three Gorges region was referred to as the *bu san bu si*, neither three nor four, meaning a place that belonged nowhere, neither to the defunct Three Gorges province nor to Sichuan province. (*Sichuan* in Chinese means "four rivers," thus the "four.") Every county here had its stories of poverty. Sandouping, chosen as the dam site, supposedly got its name, which means "three heaps of something small and round," like kernels or pebbles, from the long years when families could produce no more than three tiny piles of corn each harvest, barely enough to keep them alive. In Yunyang, the poorest county in what was then Wanxian prefecture, the tuberculosis rate was five times the national average in the mid-1980s. Fuling was too poor to pave its roads. Natives of Zigui, the poorest county in Xiling Gorge, say it was famous for two reasons: Qu Yuan, the scholar-official and poet who drowned himself nearby, and the endless lack of grain, money, and educated people. Between 1949 and 1985 Zigui received an average of 4.65 million yuan in government subsidies annually, and during these years only about ten students a year from the county were admitted to colleges, universities, teacher training institutes, or technical schools, a fact considered a particular disgrace for a town associated with one of China's most revered scholars.[4] In 1985, Li Boning, then the party secretary-designate of the Three Gorges province, made an extensive inspection tour of the area. Shocked by conditions, he made a videotape entitled "A Call from a Poor Mountainous Region." His attempts to send the video to senior party officials were initially blocked by other cadres who feared that the sight of undernourished families so poor that they owned no property was an inappropriately negative depiction of the area. Li eventually succeeded in sending copies to all party and State Council leaders. Few had any idea of the problems that existed in an area still thought of primarily in terms of its history and beauty.[5]

While China's leaders continued to debate when and at what height the dam would be built, and how the surrounding region would be administered, numerous experimental enterprises were set up with the goal of providing future employment for local people facing relocation. These ranged from a cosmetics factory in Wanxian founded with a 300,000-yuan loan from the Migrants Bureau (five years later the plant employed 302 workers and had a total production value of 15 million yuan per year) to a successful silk factory set up on a mountaintop in Fengjie with government loans and contributions from local farmers.[6] Other projects were failures, often the result of miscalculating the potential local market or how to reach a more distant one. The Fengjie county government, for example, tried in 1986 to develop a milk industry and invested 180,000 yuan in forty cows for eighteen families moving to higher ground. Although they successfully raised and milked the herd, there was no local demand for milk, and the project was dropped several years later when a delegation of the farmers poured hundreds of gallons of cows' milk into the street in front of the Fengjie Migrants Bureau.[7]

After plans for the new province were dropped, a new organization, the Three Gorges Economic Development Office, was formed to encourage and oversee the economic growth and revitalization of the area, with the Three Gorges Dam as its focal point once more. Despite this forward momentum, in June 1986 the Chinese People's Political Consultative Conference (CPPCC) issued a report opposing construction of the dam in the "short term" and raising concerns about sedimentation, flood control, and other issues. The State Council then called for a reexamination and postponement of the Three Gorges Dam project, urging that a final decision be made in a "democratic and scientific" way, and establishing a step-by-step approval process. This required the Ministry of Water Conservancy and Electric Power and the Yangtze Valley Planning Office (YVPO) to complete a new feasibility study, to be examined by the State Council and then approved by both it and the Politburo. The final step would be a vote by the National People's Congress (NPC).

In the late 1980s a growing political openness in China made possible intense domestic discussion about the Three Gorges Dam at the same time as international opinion about huge dams was becoming more negative. The report resulting from the Canadian International Development Agency feasibility study of the dam, which concluded that the project was technically, environmentally, and economically feasible, became available in China in early 1989, spurring further public debate and a number of critical essays in the press. When opponents of the dam realized that the State Council would probably approve the proposal to build, and that it could go to an NPC vote as early as March 1989, Dai Qing, a journalist and environmental activist, published a now-famous collection of essays called *Yangtze! Yangtze!* Released in the spring of 1989, the book includes mostly technical articles voicing concerns about the dam's construction, flood control abilities, sedimentation, and other problems. On April 3, 1989, the government announced that the NPC vote would be postponed and that work would not begin on the dam in the next five years, but this period of free discussion ended abruptly in June 1989 with the Tiananmen crackdown. Dai Qing was jailed for ten months in Qincheng Prison, where Li Rui had been held two decades earlier, for her outspoken opposition to the dam. Other critics were silenced by the threat of political repercussions or imprisonment.

A year later, in July 1990, at Premier Li Peng's instigation, the State Council formed a commission to review the Three Gorges project once more. This commission unanimously endorsed a 607-foot dam on August 1, 1991, and the decision to launch the Three Gorges Dam was formally approved by the State Council and Politburo in January 1992. All that remained was the approval of the National People's Congress, normally merely a rubber stamp for decisions already made by senior government leaders. On April 3, 1992, the Fifth Plenary Session of the Seventh National People's Congress approved a resolution to proceed with the Three Gorges project with a vote of 1,767 for, 177 against, and 664 abstaining. Though the numbers might suggest an overwhelming endorsement of the project, they actually revealed an extraordinary display of unease and ambivalence. The resolution won by the smallest margin of any vote in the history of the NPC. Though the results of the NPC vote may have been unpopular elsewhere, the decision to go ahead was a source of tremendous relief and enthusiasm in Yichang and Sandouping. Throughout the afternoon, every household or shop that had a television had it on full blast so that everyone in the neighborhood could hear. In Sandouping township, residents hung hundreds of scrolls and banners in the streets declaring their hopes for the Three Gorges Dam, and more money was spent on firecrackers that day than for the New Year's celebrations.

As soon as the resolution was passed, hundreds of fireworks exploded, and, according to news reports, people wept in the streets. "When an old worker of the Three Gorges exploration team heard the news, he slowly raised his head, and with tears rolling down his cheeks, declared loudly, 'I have been working here for almost thirty years, but it has been worth my sweat and tears. I am crying with joy today.'"[8] For idle state-employed construction workers from the Gezhouba Dam, who had been kept on the payroll at 70 percent of their salaries since completion of that dam, approval of the Three Gorges project meant hope of a transfer to the new site and the possibility of a full income. Denied both permission and the opportunity to move on to other work for almost five years, over 50,000 men and their families had been living in squalid conditions near Yichang, in increasingly difficult circumstances. When the decision was announced, thousands of men laden with liquor and firecrackers celebrated on the streets.

In Yichang and the Hubei sections of the gorges, the press was filled with stories of people who had sacrificed for the dam. Wang Yu, a twelve-year-old boy from Zigui had learned only the year before about the construction of the Three Gorges Dam. Recognizing that the state was too poor to build the dam without assistance, he decided to donate his entire 10-yuan savings to help.[9] On a grander scale, officials from Yichang began collecting funds months in advance in order to make a 100,000-yuan donation to express the support of the people of the Three Gorges for the dam. Immediately after the vote, the sum, huge by local standards, was turned over to the Three Gorges Economic Development Office of the State Council.[10]

In Sandouping, the future site of the dam in Xiling Gorge, and Zhongbao Island, which would become the base of the dam, life changed immediately. The only island within the gorges, Zhongbao was gradually destroyed as the dam was built around and on top of it, but for a few years before that, it attained a brief period of fame as tourists and world leaders stopped to see the piece of land that would one day be at the bottom of the biggest dam in the world. Given the dangers of Xiling Gorge, the advantages of living on the island do not immediately come to mind, but it had enjoyed an unusual prosperity because of its good soil and small population. Zhongbao (meaning "central treasure") was believed by local people to have once been a golden bowl with gold lions on its handles, guarded by ducks and horses that protected it from flood. Zhongbao Island was inhabited by members of the Neolithic Daxi culture over 7,000 years ago, and more than 5,000 artifacts were found there during preparations for the Gezhouba and Three Gorges dams. Since the 1950s, when investigative work began on the dam site, stories have been told about the residents of the island, members of the Sandouping Commune, who observed or took part in preparatory work for the Three Gorges Dam. Gao Sanyuan, an old man from Zhongbao Island, is honored in one local tale for his loyalty in guarding the first test well drilled on the island to sample the granite base.[11] Initially believing that the noise of the drills would anger the dragon king and cause a flood, Gao eventually came to grasp the importance of the engineers' work, if not the exact purpose of the well, which he kept protected from rain and cold for over thirty years after the drilling stopped. Only when the NPC resolution to build the Three Gorges Dam was passed did he cease his work.

After the vote, villagers on Zhongbao Island put up new markers and signposts for the dam. A Mrs. Huang charged two mao (twenty Chinese cents, or about four U.S. cents at the time) for a photo of her field. Nearby, Mr. Zhao, made 600 yuan per day from an average of 3,000 people stopping to take pictures of the pole he had erected on his plot of land to mark the site of the Three Gorges Dam.[12] In May 1992 Xinhua News Agency reported that more than 20,000 tourists were visiting Zhongbao Island daily and that accommodations in Yichang were filled.[13] The village looked much like every other rural village in the area, but with its posts and markers symbolizing a dam that would bring triumph to China, it was, for a short time, an historic site. Within two years, the people were gone and the island, with its solid granite core, was invisible beneath the cranes and dredges.

CHAPTER 10

THE DAM

THE GROUNDBREAKING CEREMONY FOR THE Three Gorges Dam was held December 14, 1994. After years of discussion and postponement, actually digging a hole in the ground for the dam meant far more than all the approvals and votes and study papers of the past forty years. During the summer and fall of 1994, thousands of peasants from Sandouping left their homes because of the Three Gorges Dam and went off to new villages and towns and factory jobs, and early in 1995, iron and steel bridges and gates, concrete walls and deep passageways began to emerge out of the gargantuan holes and dirt piles. A seventeen-mile road to Yichang, with seven miles of bridges and tunnels and a price tag of U.S. $110 million, cut through the mountains of Xiling Gorge. Yellow and red bulldozers and jeeps rumbled back and forth, still looking odd and out of place to people sailing by below. From opposite banks, the two ends of what would become the first bridge in the gorges stretched out to reach one another. At 2,950 feet, it is the longest suspension bridge in China, except for the Tsing Ma airport bridge connecting Hong Kong to Lantau Island, completed in 1997.

Three years after the groundbreaking, Chinese President Jiang Zemin and Premier Li Peng took part in the celebration to mark the damming of the river and the opening of a diversion channel that would permit river traffic to continue without interruption during construction on the riverbed. The diversion of the river, which ended the first phase of construction of the project and allowed work on the actual dam site to begin, had originally been planned for 1998, but the date was moved up so that this event and Hong

Kong's return to China could take place in the same year. A local publication described the event:

> On November 8th, 1997, a slight fog filled the Xiling Gorge and a warm wind blew gently. At nine o'clock, Premier Li Peng ordered the start of the closure at the work site near the dragon's mouth. Three signal flares sent up by the vice-general rose to the sky, and instantly trucks on the four embankments of the upper and lower coffer dams approached the dragon's mouth like powerful lions. More than 4,000 giant loading trucks poured stones into the dragon's mouth, like thunder. The dragon's mouth was narrowed to 30m, 20m, 15m, 5m. . . . The dam site had a grand, festive atmosphere. An enormous five-star red flag made up of woven blankets covered a space of 1,080 square meters and looked magnificent from the reviewing stand.[1]

In his speech that afternoon, President Jiang Zemin emphasized the benefits of the dam and extolled China's long and successful history of combating nature.

> Since the dawn of history, the Chinese nation has been engaged in the great feat of conquering, developing, and exploiting nature. The legend of the mythical bird Jingwei determined to fill the sea with pebbles, the Foolish Old Man resolved to move the mountains standing in his way, and the tale of the Great Yu who harnessed the Great Floods are just some of the examples of the Chinese people's indomitable spirit in successfully conquering nature. The scale and overall benefits of the water conservancy and hydropower project we are building today on the Three Gorges of the Yangtze River, which have no parallel in the world, will greatly promote the development of our national economy and prove to be of lasting service to present and future generations. It also embodies the great industrious and dauntless spirit of the Chinese nation and displays the daring vision of the Chinese people for new horizons and a better future in the course of their reform and opening up.[2]

If you ask people in the gorges what they think about the dam, the most common first answer is "it will be very big!" (*hao da!*). If you wonder what the area will be like once the dam is finished or how life will change, responses vary but reflect the widespread belief that most people in the cities will be better off, and that many in the countryside will not. Whatever does happen, and expectations for the future tend to be vague here, everyone agrees that it will be out of their control. Though where everyone in the way of the water will go and how this happens is crucial to the dam's success or failure, many young people project a deep indifference. "We'll all be gone," says a young woman in her early twenties, sucking on a purple popsicle outside a shop in Shibao Block, "and it doesn't matter." "It will be fun when we all move" states a chambermaid from the Wushan Hotel with enthusiasm. Pressed to explain why, she giggles and says everything will be new. The opinion of a local schoolteacher traveling third class on a local ferry is that the dam will cause widespread hardship but is necessary to bring economic development. He sees no other choice. "The people are bitterly poor and we have to do something." An assistant dockmaster in Zigui is more fatalistic and fed up. "Who knows what the dam will do in the long run, or what will become of the people. It's the cadres who make the decisions and take the money and work things out to suit themselves." The refrain from the local travel service is consistently and officially positive. Despite the tremendous cultural loss, the local offices of the China International Travel Service tout the enhanced scenery that will result when the water level is raised. A big lake will be more beautiful, and once-inaccessible spots will become easy-to-reach tourist destinations on newly created rivers.

The Three Gorges Dam will certainly be big. At a height of 607 feet, it will be about as tall as a 60-story building, and stretch 1.45 miles across the Yangtze, five times as long as the Hoover Dam. Twenty-six huge turbines will generate 18.2 million kilowatts of energy, comparable to the output of eighteen nuclear power plants and eight times greater than that of the Aswan Dam in Egypt. The reservoir water level will rise to 574 feet, with an average increase in the water level within the gorges of about 290 feet, creating a 360-mile reservoir stretching from the dam site to Chongqing. About the length of Lake Superior, this will be the largest man-made body of water in the world, with a capacity of 11 trillion gallons. Once the Wuhan Bridge is somehow altered to allow taller ships to pass underneath it, 10,000-ton vessels will be able to sail through twin 65 foot-high five-stage locks, assisted in their passage by the highest ship elevator in the world. Construction of the dam will require over 10 billion pounds of cement, 4.24 billion pounds of rolled steel, and over 56 million cubic feet of timber. Two billion cubic feet of rock were blasted away to carve out space for the massive locks alone; 240 square miles of land will be flooded, submerging at least thirteen cities, 140 towns, 1,350 villages, 657 factories, and approximately 74,000 acres of cultivated land under about 300 feet of water. Some 1.2 million people may eventually be moved away from those towns, factories, and farms.

Although each of the Three Gorges will be affected by the new dam, as will the towns and countryside as far away as Chongqing, no place has or will suffer greater change than Xiling Gorge, the easternmost part of which will be

Figure 10.1 Poster at the Three Gorges Dam site: "The Three Gorges police are always prepared to maintain security"

totally destroyed. In the fifth century, Li Daoyuan, a scholar of the Northern Wei dynasty wrote,

> About fifty kilometers east of the Yellow Cow shoal is the mouth of the Xiling Gorge. Here, the landscape features high peaks and winding waterways, the flanking mountains being so high that sunshine can filter in only at midday and moonlight can come only at midnight. The precipitous cliffs tower thousands of feet and bear colorful streaks with myriad images. Ancient and tall trees abound. Apart from the singing streams, one can hear clearly the echoing wails of monkeys inhabiting the mountains.[3]

For 1,500 years, this remained almost unchanged, and in the space of a decade it has become unrecognizable.

The numbers of tons of earth do not begin to convey the enormity of the construction and destruction, the walls of concrete as high as skyscrapers in the midst of what will soon be remembered only from old photos. The sounds are different too. The monkeys are long gone, and so are the shouts of the boatmen, for it has become too dangerous for sampans and other small craft to maneuver around the dredges and cranes, and the gurgling of the river has been drowned out by the rhythmic slam of pressure drills. On one side of the riverbank, slogans in huge red characters declare, *Yiliu guanli, yiliu zhiliang, yiliu shigong, wenming jianshe* (Top-rate management, top-rate quality, top-rate workmanship, and civilized construction), on the other, *Kaifa sanxia, fazhan changjiang!* (Open the Three Gorges, develop the Yangtze!).[4]

Worldwide, there are now over one hundred dams with heights over one hundred meters (equal to 328 feet). Their reservoirs cover more than 230,000 square miles, and have a total water capacity of 212 billion cubic feet, equivalent to 15 percent of the annual runoff of the world's rivers. With the exception of the Zaire, the Amazon, and rivers flowing into the Arctic, all of the earth's thirty largest rivers have been dammed; these include the Ganges, Panama, Tocantis, Columbia, Zambezi, Niger, Danube, Nile, and Indus.[5] The Three Gorges Dam will generate 50 percent more power than the Itaipu Dam in Paraguay, currently the world's largest dam.[6] Despite a worldwide trend against building mega-dams, a consequence of the environmental damage and social problems resulting from many of the projects listed above, the Chinese government views the Three Gorges Dam as the most effective method of energy generation and flood control along the Yangtze, as well as a way to improve transportation to the interior of China and jump-start the economy in one of the country's poorest regions.

By 2009, the dam should provide 10 percent of China's total electric power supply, but most of the newly generated power will go to the eastern cities and provinces of the Yangtze River delta. Power originally designated for Chongqing will instead be sold to Guangzhou, and any deficiency in Chongqing will be made up by the Ertan Dam in southwestern Sichuan. The Three Gorges region will receive little of the electricity and, at least for the time being, has little need of it, for there has been a power glut in most of Sichuan since the mid-1990s, the result of increased production by small generating stations and the closure of many large state-run factories. Nonetheless, blackouts and brownouts are still common, mainly because of unequal distribution rather than lack of supply. Even if electricity from the Three Gorges Dam were to become available locally, it would not necessarily be welcome. Many county officials do not want electricity from the dam, for they fear it will result in the closure of local power-generating stations, and consequently the loss of tax revenue, greater unemployment, and generally higher electricity rates.

Though some foreign calculations run up to U.S. $30 billion and higher, in early 2001 the Chinese government put the anticipated cost of the dam at 180 billion yuan (U.S. $21.74 billion), almost all of which is coming from domestic sources. This is 23.9 billion yuan (U.S. $2.89 billion) less than the 203.9 billion yuan (U.S. $24.63 billion) predicted in 1994. One hundred billion yuan, or close to 50 percent of the original estimate, is being raised directly from a .7 fen (.08 U.S. cents) per-kilowatt tax levied on electricity on all homes and businesses in developed regions of China. (Less-developed areas are charged .3 fen and the poorest counties are exempt.) Revenue from the Gezhouba Dam is providing 25 percent of the funding. A ten-year, 30 billion yuan (U.S. $3.6 billion) loan was granted by the China Development Bank, established in 1996 to fund state-owned enterprises and major infrastructure projects, including, in particular, the Three Gorges Dam. The remainder of the financing comes from export credits and corporate bonds, some of which have been purchased by U.S. financial institutions despite significant opposition and pressure from American environmental activists. In 2003, when the first turbines begin generating energy, the Three Gorges project will begin contributing to its own support and will eventually pay for about 7.5 percent of the total cost. While China had originally hoped for outside financing, the World Bank and other international financial institutions declined to provide assistance because of concerns about the environmental impact of the dam. Financial investment from the United States has been limited compared to that from Europe and Asia because of a 1996 decision by the Export-Import Bank of the United States not to offer credit guarantees (generally required

Figure 10.2 Three Gorges Dam site

by China) to American companies wishing to provide equipment or services to the Three Gorges project. Despite this, U.S. companies, either acting directly or through overseas subsidiaries, have still managed to do over U.S. $100 million worth of business related to the project, and there has been no shortage of interest from major engineering and construction companies throughout the world.

In part to simplify the administration of central government and other funds coming into the Sichuan side of the gorges, as well as to coordinate the population relocations and infrastructure development, in 1997 the city of Chongqing was elevated to the level of a municipality, like Beijing, Tianjin, and Shanghai, and enlarged to include the counties of the Three Gorges as far east as the Hubei border. Chongqing municipality, no longer a part of Sichuan province, now reports directly to the central government. Wanxian, which had administrative responsibility for most of the Three Gorges counties from 1992 to 1997, was renamed Wanzhou, or Wan district (though many people still call it Wanxian), in 1998. Also referred to as the Wanzhou Resettlement Development District (*Wanzhou Yimin Kaifa Qu*), it includes Fengjie, Wushan, Wuxi, Yunyang, Kaixian, and Zhongxian counties, and reports to the Chongqing government.

The biggest question about the dam is whether or not it will be successful in reducing flooding in the middle reaches of the river, where most of the worst floods occur and the lives and livelihoods of hundreds of thousands of people living behind the Jingzhou Dike on the flat farmland of Hubei are threatened. Such floods happen during the summer rainy season when there is a sudden infusion of water from tributaries, usually the Han and Huai rivers near Wuhan, and the Yangtze is already rising rapidly as melting snow swells its uppermost regions. The Yangtze flows from west to east. In late spring and summer, rainfall along the river moves in the opposite direction, from east to west. Heavy rains often fall in central China, in Hubei and Sichuan, in midsummer, coinciding with the time when the Yangtze is already at its highest. Several situations can lead to catastrophic floods: excess water entering the Yangtze from its tributaries, too much water flowing from the upper Yangtze above Chongqing into the gorges and the middle reaches of the river below, or a combination of these circumstances. While concurring that the Three Gorges Dam will have some impact on flooding by limiting the flow out of the reservoir area, and thus reducing the overall volume of water below Yichang, many experts argue that a dam in this location can only be effective in preventing disasters caused by flooding in the upper reaches. The flood of 1870, considered a "thousand-year flood" (a flood which is so rare that statistically it should occur only once every thousand years), is an example of what happens when there is far too much water pouring into both the upper Yangtze near Chongqing and the gorges and the middle Yangtze (from the tributaries) simultaneously. A dam located in Sandouping might have reduced the damage caused by a flood of this sort by containing the water from the upper reaches of the river in the reservoir.

The hundred-year floods of 1954 and 1998, in contrast, were the result of water pouring into the Yangtze from tributaries below Yichang. Critics claim that when the tributaries are the source of the flooding, lowering the reservoir level above the tributaries, in anticipation of the increased volume of water upriver, will not significantly improve the situation, certainly not enough to justify the building of a huge dam with so many other problems and potentially negative consequences. Supporters claim that it will. A reservoir provides flood control by working like a sink, allowing water that enters it to be drained at a controlled rate, rather than overflowing unexpectedly. Before the rainy season, the reservoir level can be lowered, providing additional storage capacity that will prevent flooding in this region and below the dam. If, however, the increase in the water level below the dam from the tributaries in the middle reaches of the Yangtze is so great as to cause flooding, a dam above may reduce the scope the disaster, but it will not prevent it.

Long-term opponents of the Three Gorges Dam, such as Li Rui and Huang Wanli, for decades advocated the construction of smaller dams along the tributaries as a potentially more effective method of flood control in this area. In general, international policy toward dam-building has shifted in this direction as well. Although numerous dams have been built on the tributaries of the Yangtze for irrigation, power, and regional flood control, China has long dismissed the idea of using multiple small dams as a substitute for the Three Gorges Dam, claiming that it would be an inefficient solution for large-scale flood control and hydroelectric generation.

As recently as 1998, the death toll along the Yangtze reached close to 4,000 in the summer floods. Another 2,000 people died in floods there in 1991. During the twentieth century alone close to half a million people were killed by flooding along the Yangtze. Government assurances that the dam will reduce this threat make a convincing enough argument for why the dam must be built for most people in the gorges, even if it will not benefit them personally, and few have ever heard of any controversy about whether it will work. The slogan "The Three Gorges belongs to the nation" is plastered everywhere, and banners strung across the streets in every river town emphasize the necessary sacrifice. Despite the damage to the environment and the difficulties of relocation, most residents take for granted that government engineers know how to build a dam that will do what it is supposed to do. In reality, this has not always been the case. Of the approximately 80,000 dams built in People's Republic of China during its first forty years, an average of 110 dams collapsed each year between 1950 and 1980, including a record 554 collapses in 1973 alone. One of the worst disasters in China's history was the 1975 collapse of sixty-two dams in Henan province in one night of terrible typhoons.[7] In this tragedy, the Shimantan and Banqiao dams, two supposedly indestructible "iron" dams, built in the early 1950s on a Soviet model, burst after just two days of wind and rain, drowning somewhere between 86,000 and 230,000 people, depending on whether one uses the official figures or what is believed to be a more accurate estimate. Another eleven million flood victims suffered from disease and famine for months thereafter.[8]

Though the majority of the dams that collapsed were built in haste and often by workers with no technical training during the Great Leap Forward and the Cultural Revolution, the idea of the collapse of a structure the size of the Three Gorges Dam remains terrifying. After several serious accidents, an inspection in 1999 revealed problems in construction of one wall of the dam, which was dismantled and rebuilt, and led to a crackdown on safety violations and the use of inferior materials. In December of the same year, Premier Zhu Rongji announced that the central government must thoroughly investigate all suspected cases of embezzlement and improve financial management. Two hundred officials were assigned to work full-time on investigating corruption related to construction of the dam, and 105 officials associated with the project were arrested in January 2000 for extortion and taking kickbacks. Cracks in the dam detected between 1999 and 2002 raised further concerns. A break in the dam, whether the result of poor construction, an earthquake, or sabotage, would release 10.8 million cubic feet of water per second and cause a flood forty times that resulting from the destruction of all sixty-two dams in 1975.

Assuming nothing so dramatic occurs, there are still other things to worry about. The reduction in the rate of flow of the Yangtze is expected to have serious consequences, one being the enormous buildup of silt in Chongqing harbor as a portion of the almost 585,000 tons of silt now carried out by the river and deposited in Shanghai settles instead at the end of the reservoir. By 1964, four years after the Sanmenxia Dam on the Yellow River was completed, five billion tons of sand had collected, clogging the dam and threatening the city of Xi'an. The reservoir had to be emptied, dredged, and rebuilt, and ended up with less than half its original capacity. Given the potential problems, Chinese scientists and officials continue to consider the possibility of keeping the summer reservoir level significantly lower than its full normal level of 574 feet. The height of the dam, and thus the water level if the reservoir is filled to capacity, are directly related to the increase in sedimentation at the other end of the reservoir: in general, a higher dam and deeper reservoir will result in greater sedimentation. If, as some experts expect will be the case, the reservoir is lowered to 476 feet (the anticipated level of the reservoir in 2006) during the summer months, the accumulation of sand in Chongqing will be reduced.

Pollution presents other problems. A decrease in the rate of flow of the river may also lead to significant problems for human and animal health as huge amounts of sewage, chemicals, and other contaminants that in the past were flushed out by the fast-flowing river will now collect. Schistosomiasis, a parasitic disease closely associated with the development of liver cancer, is widespread in rice-growing areas of China and is persistent along the Yangtze despite decades of effort to eradicate it. The creation of a large, slow-moving body of water is expected to make this worse. Over 3,000 factories and quarries discharge an estimated one billion tons of waste water annually, along with over 8,800 tons of toxic pollutants of fifty identifiable types, including large amounts of lead and mercury. Additional heavy metals may leach into the water as defunct factories are submerged without adequate decontamination. The central city of Chongqing alone pours almost a billion tons of waste water into the river every year, and 80 percent of all sewage there and elsewhere along the river is untreated. An estimated three million tons of garbage in 605 dumpsites between Chongqing and Yichang will be under water after the flood, and although the Chongqing municipal government hopes to dispose of all this before the water rises to 443 feet in 2003, the cost is estimated at 3 billion yuan, of which so far only 463 million yuan has been allocated.[9] Trash collection on the Yangtze itself is far better than it was a decade or so ago, and large ships now empty thousands of pounds of garbage onto special barges in Wuhan and Chongqing every week, rather than tossing it into the water, but this still ends up in the dumps on the shore. For decades, the custom of giving anything unwanted to the river to eat has left the Yangtze filled with items it should never have been fed. Plastic bags, bubbling factory foams, furniture, machinery, and waste of every imaginable sort hurtle down the river or are dragged unseen along the bottom. Once the river becomes a reservoir, this and much else will have to change. If not, the reservoir has the potential of becoming a large and filthy lake.

The increase in water volume may also affect the climate, something already seen in China in Yunnan province, where the reduction in the size of Lake Dianchi during the Cultural Revolution has made winters colder and summers warmer. The increased volume of water in the gorges is expected to have a moderating influence on the temperature, lowering it by .9 to 1.8 degrees Fahrenheit in summer, and increasing it by 1.4 to 2.3 degrees in the winter. Such a change is considered a potential threat to agriculture, particularly to tangerines, which have been one of the region's main cash crops since the 1970s and grow only at a very narrow band of elevation and temperature. Even without worries about the temperature, over 12,350 acres of tangerine orchards and 95 percent of the longan and lychee trees will be submerged. The possible environmental payoff is that the dam, if successful, will produce enough clean energy (the equivalent of burning 55 million tons of coal a year) to reduce emission rates of carbon dioxide and sulphur dioxide, pollutants that contribute greatly to China's terrible air quality and to climate changes.

The Three Gorges region is also home to a large number of rare plants, many of which are ingredients in traditional Chinese medicines. Some, such as the lotus-leaved brake used to treat kidney disease, are found nowhere else in the world. Forty-seven plants, of which thirty-six are found only in the Three Gorges, are considered rare and face extinction. Chen Weilie, a researcher at the Institute of Botany of the Chinese Academy of Sciences, said in 1994 that land reclamation had already gone beyond the endurance of the local ecology—and that was before thousands of acres of fertile land were to be inundated.[10] Botanists are attempting to preserve these rare plants and protect other plant species by setting up three national protection zones in the Shennongjia region of Hubei province. One protected area, at 7,200 feet, is for the conservation of primeval forest. The other two, at 4,300 feet and 1,300 to 2,600 feet, are for the study of plants endangered by agriculture and urban development. About 40 percent of the land in the Three Gorges was under cultivation in the year 2000, already putting its natural resources under tremendous pressure, and the forested area, which once covered most of the mountainsides, had dropped from 20 percent in the 1950s to less than 10 percent. A forest cover of 30 to 40 percent is considered necessary to have a significant impact on erosion and flooding, and was set as a goal by the government after the disastrous 1998 floods.

Overall, wildlife of any sort has not fared well along the river either. Ancient poetry describes the howling of apes and chattering of birds, and though a few monkeys live on the cliffs of the Daning River and a few other tributaries, for the most part animal life now can only be imagined. Most of the Yangtze River sturgeon, the huge fish that Mr. Tyler's workmen took for a dragon while blasting in Xiling Gorge, met their demise with the completion of the Gezhouba Dam. Although a fish ladder was built to allow them to swim upriver to spawn, the fish, as a local official explained, were not smart enough to learn how to use it. The pink Yangtze River dolphin is another creature mentioned in thousand-year-old poems but not often spotted today. A natural preserve has been set up in an estuary of the Yangtze in Hubei province, below the dam, where the water level will not change, and a few sturgeon and porpoises now are the lonely occupants of an aquarium in Yichang. There is, however, still little interest in this issue locally, where animals, no matter how rare, are largely regarded as something to eat. Efforts to breed the sturgeon have met with some success, but the fish still fail in the attempt to swim upriver, often exhausting and injuring themselves in the process. Environmentalists fear that, with the reduction in the rate of the water flow once the river is dammed and this section becomes a reservoir, the fish are in for an even harder time as silt, sewage, and other toxins that were previously washed away by the water collect. A number of other animal species are also endangered by changes to their environment, including close to eighty species of fish, the finless porpoise, the Yangtze alligator, the Siberian white crane, Temminick's cat, and the South China Siku deer.

The people and animals living along the shores of the Yangtze are not the only ones affected by the rising flood waters. Thousands of graves are being flooded despite the deeply rooted belief in *feng shui*, whereby physical objects, including houses and graves, must be put in the right alignment with nature. Most peasants here continue to believe that the location of their parents' and ancestors' bones is of utmost importance in determining the welfare of their descendants. Disturbing the remains of the dead can lead to disaster for the living. The National Cultural Relics Bureau has identified 1,208 sites and buildings considered of key significance, but little can be done to save most of them. Four archeological excavation sites connected with Ba society as well as the earliest Neolithic site in the Three Gorges are included on this list. Since plans for flooding began, Yu Weichao, director of the Museum of Chinese History in Beijing and in charge of cultural preservation for the Three Gorges region, has petitioned for additional funds and manpower for the preservation of archeological and cultural sites along the river, but to no avail. There is simply not enough time or money, and the loss is not only of what is known to exist, but of what recent archeological finds suggest is yet to be discovered.

Numerous valuable cultural relics have been unearthed during the construction process, offering antique dealers and looters new and extraordinary opportunities. Sophisticated thieves equipped with cell phones, high-frequency radios, and prospecting equipment have stolen hundreds of items from the gorges, particularly in Wushan and Fengjie counties, where excavations were stalled in 1999 and 2000 because of a lack of funds. The original budget of nearly two billion yuan (U.S. $243 million) for excavation and preservation was reduced to 300 million yuan (U.S. $36.3 million) and there have been prolonged difficulties in the distribution of the funds to local authorities because of conflicts about control of the money. One of the more impressive feats of thievery was exposed in March 1998, when a four-foot Han dynasty candelabrum, believed to be from Baidicheng, near Fengjie, sold for U.S. $2.5 million at the International Asian Art Fair in New York.[11] The piece, called a spirit tree, is one of only three known to exist in China. Dealers in Hong Kong say there is a constant trickle of items from the region. The thief was reportedly caught and imprisoned in Shaanxi but managed to bribe a jailer and escape.

In addition to issues related to flood control, the environment, and cultural preservation, international groups have also raised concerns about potential human rights abuses in the use of prison labor on the dam and in the forced relocation of villagers throughout the region. In addition to paid construction workers, thousands of prisoners from jails in Hubei province reportedly work on the site. I asked a young tour guide showing people around the dam site if she had heard about this. Agitated, she responded that it could not be true that prisoners were working on the dam. China would never allow convicts to take part in such a glorious project.

Most difficult to deal with is the disruption to people's lives and the uncertainty of the future. In the mid-1980s, Chinese authorities estimated that approximately one million people would have to be relocated because of the dam. Many officials now put the number at closer to 1.2 million, once new population figures and children expected to be born by 2009 were taken into account, but no one is entirely sure what the final figure will be. The problem of how to find housing for people who do not yet exist is overwhelming. Nowhere is the divide between those who are willing to move and those who want to stay clearer than between urban and rural residents. As most people say, life will generally be more pleasant for city residents after the move, but for many peasants, it can only get harder in the years to come.

The cities along the Yangtze between Chongqing and Yichang, for the most part, are an unpleasant mix of a few surviving old and crumbling wood or stone buildings interspersed with the even more unpleasant concrete blocks of the past fifty years. Indoor plumbing is still a luxury for many in all but the biggest cities. In Shibao Block, for instance, the government headquarters and apartment buildings for municipal employees, built in the mid-1980s, had one squalid outdoor latrine for both the offices and the twenty apartments. Though some of the occupants rigged up showers on their balconies, most made do with the male or female "washing rooms," concrete chambers across from the outhouse, each with one cold water tap and a bucket, but large enough for a dozen people. Utilitarian Soviet-style housing was built for shelter, not comfort.

The new cities offer new buildings and apartment houses that are not grim and grimy—at least not yet—something almost no one living in these towns has ever known. Considering that they were built extremely quickly and as cheaply as possible, it is hard to say how they will hold up a decade later, but for now they represent modernization and choice. The disadvantage is that housing is no longer inexpensive. Residents must buy their own apartments now that the days of government subsidies are over, but years of rents set at a tiny fraction of their incomes have made it possible, though sometimes difficult, for most people who are employed by government organizations to do so. Those who cannot afford their apartment of choice must usually crowd into smaller, less expensive spaces. Although this is not a novel dilemma in most other places, in a country where for two generations, regardless of position or income, all but the highest ranking officials lived in cramped and uncomfortable apartments, the sudden and wholesale destruction of the cities along the river means facing a new kind of inequity.

The per-square-foot price in an average building in a town like Fengjie or Wushan can be more than 45 yuan (compared to an average of 372 yuan a square foot in the most expensive parts of Shanghai, or about 75 yuan in a new residential section of Chongqing), but compensation for apartments in the old river cities, in contrast, is rarely more than 35 yuan per square foot. In the countryside it is less, even if the people must move to an urban township. The discrepancy between the cost of new housing and compensation for the old causes severe hardships, especially if workers have been laid off or have had to cope with expensive illnesses or other problems. The least expensive and smallest apartments start at about 20,000 yuan for two to three small rooms and go up from there, with a comfortable three-bedroom apartment in Fengdu or Wanxian costing about 50,000 yuan, depending on the location and type of building. Beyond this is the expense of outfitting the apartment. As is common in much of Asia, the buyer must obtain and install all the plumbing and fittings and finish poured concrete floors and walls. For many urban residents, who have lived in quarters assigned by their employers for their entire lives, this is the first time they have the opportunity to choose where they will live or to spend more for something they want, and they are enjoying it. On the streets of Wanxian and Fengjie there are sections of small shops filled with interior piping, washing machines, linoleum, and wallpaper, none of which existed five years earlier. "Business is good," one of the owners tells me, "everyone has to move." At the higher end of the market are stores that sell matching drapes and couches and offer advice on interior design.

Despite this relative freedom, the majority of city residents will move to a building constructed for or by their *danwei*, or work-unit, and will live with neighbors from their school, hospital, municipal office, or wherever else they work and which has always provided their housing. Independently employed city dwellers, a group that consists mostly of shopkeepers, medical practitioners, and small business owners, are likewise usually grouped together, their moves coordinated by a special office. In addition, most towns also have some luxury housing for high-level cadres, for example, or expensive apartments available to wealthy entrepreneurs, but there is not much of this. In general, choices are still determined by one's workplace.

Even with the general satisfaction with the new housing, many city residents are frustrated at being forced to move to half-built cities where services are lacking, or are angry when they discover that they are in buildings that have already begun to deteriorate while government officials head to more opulent accommodations. Others are anxious because relocation subsidies will not cover the cost of moving their furniture and are dismayed by the chaos and problems of transportation, with half the city already in new housing on a mountaintop and the other half still below waiting to be washed away. In most places, people will miss the familiar streets and sense of connection with the past, but, overall, housing has been so awful for so long that, though the residents grumble, few are truly sorry to go.

Hope for the future goes beyond just new apartments to include better schools, new businesses, and outside investment. Only with a more diverse and modernized economy, with factories that produce high-quality goods that someone wants, and with fewer people depending on agriculture for their livelihood, can the river towns hope to prosper. New roads and the ongoing reorganization of the many old state-run enterprises in the region may help, but it will take more than this. As one former official in Fengjie said to me, the problem is not just the Three Gorges, it is all of China. The social and economic problems that are commonplace are exacerbated by the dam project and the relocation, but are shared by cities and towns everywhere now that the large state-owned enterprises that supported much of the urban workforce are being shut down. In the gorges, the enormous amount of work and energy required to build new cities, relocate the population, and put functioning medical and educational facilities in place act as a temporary diversion from facing the long-term questions about the future development of the region. Everyone is too busy just coping with the present.

This can result in an odd kind of myopia, or perhaps disingenuousness. Government officials in Chongqing explain that newly created schools for migrants, like all government enterprises these days, will have to find ways to provide at least part of their own support by running adjunct industries or locating investors; yet, in the same breath, they point to the lack of interest in the region shown by outside businesses, the high unemployment, and the limited local market for most light industrial goods that could be produced by a school-run factory. How do they reconcile this need for profit-making activities with a generally unpromising economic situation? They don't know, they are not economists, but the schools will figure something out. A key point of the reservoir relocation program is the intertwining of all segments of development, and specialists are clearly intent upon making an improvement in their own fields. Despite this, in the towns there seems to be not only an odd lack of awareness of the connections between one area and another, but a lack of interest as well. It may be simply that there is too much change at once or a bureaucratic tradition that says what is not my responsibility does not concern me. Whatever the cause, you sometimes get a sense of individual units groping blindly toward the future, trusting that at some point all the parts will fall together and make a whole that works, and if it does not, it was not their responsibility anyway.

As the towns are dismantled, people make their preparations for the future in different ways. For some, it is merely packing up their belongings and picking out appliances. For others, the move brings new uncertainties and sometimes a renewed hope of leaving the city at a more fluid time than usual, when so much change may mean that the authorities are more open to persuasion. For people in professions like teaching or the government tourist service, who have often been assigned from other parts of Sichuan to towns like Wushan and Fengdu, the prospect of staying on through the relocation implies a permanence they never intended, and can spur frantic efforts to get out before they have to move uphill.

Young employees surreptitiously apply to graduate school and jobs in other cities, trying, until they have a secure plan for the future, not to let their colleagues or more senior officials know. One young official I met went to take the examinations for a graduate program in Chengdu under the ruse of visiting her mother. Concerned that the neighbors would notice any unusual mail, she had the scores sent to her parents, but anxious for the results and unwilling to wait, she went to a local fortuneteller, trusting that this would be as efficient as a phone call home. She will be gone by the time the water rises, but the majority of individuals with mid-level government jobs were born and raised in these cities and have no desire to go anywhere else. The towns will be in a slightly different location, familiar roads and homes will be gone, but their neighbors and work will be the same, just a hundred meters higher, and not too much will change, they think. Maybe, if all goes well, there will be better schools, better shops, more jobs, and faster transportation to Chongqing and Wanxian.

In the countryside, it is entirely different. Generations of peasants have lived together in adjacent villages, growing (except when interrupted by political upheaval or natural disaster) their sorghum or corn or tangerines, and taking husbands and wives from other nearby villages. Years of work have transformed steep mountainsides into a patchwork of multi-hued squares and strips of grain and vegetables. The peasants here must start over again, and except for the few who can move directly up a short distance from where they were before, the disruption is complete and wrenching. During the initial stage of the relocation, the government had planned to move almost all rural migrants, approximately 420,000 people, to higher land as close to their original homes as possible. The policy, known as *jiu di, hou kao* (meaning "move close to the old place," but back, or higher up), assumed that while some peasants would be moved into nonagricultural jobs, either in the countryside or in new townships, the majority would go to nearby sites in the same village groupings and continue to farm. Despite efforts to rehabilitate mountainsides for farming, uprooting hundreds of thousands of relatively poor villagers and moving them to poorer areas or removing peasants from profitable farms and orchards has proved difficult even for a country accustomed to the large-scale relocation of people to places they do not particularly want to go. With fertile land scarce, the heightened competition for limited resources has led to angry protests by both peasants required to move and those already living on higher ground, and organized attempts to voice concern or resist relocation have led to rapid government intervention. Officials have long been aware of the potential for an explosive situation, as this 1994 document by the Wanxian Public Security Bureau makes clear.

> In matters such as compensation for property and the allocation of land and housing sites, a gap between the relocatees' hoped-for valuation and the actual compensation is bound to occur. . . . The conflict between expectations and frustrated desires could easily foster antagonism toward the government on the part of the relocatees and lead to some of the people refusing to move. This would hamper the smooth progress of the Three Gorges project. A slight exacerbation of this antagonism, coupled with agitation by unlawful elements, will lead to disturbances such as sit-ins, demonstrations, and petitions, or even grave cases of beating, smashing, and looting.

Figure 10.3 Prosperous farmer's fields to be submerged near Fengjie

> There is no lack of precedent for such disturbances. For example, problems left over from the relocation of inhabitants in projects such as the Xinhua Reservoir in Wushan county, the Xiaojiang hydropower station in Yunyang county, and the Baishi Reservoir in Zhong county have constantly been the cause of frequent mass disturbances of no small scale; these problems have not been overcome even at this time. These small reservoirs involved the relocation of only a few hundred people, whereas the Three Gorges project will involve a million. Once disturbances arise, it is difficult to predict how far they might go.[12]

Since the mid-1990s, problems in Yunyang, one of the poorest counties to be flooded, have been described extensively by the anthropologist Jing Jun. Compensation amounts for relocation are determined by a number of factors, including whether or not a place is considered urban or rural and the estimated value of the housing or land that is being lost. In Yunyang, as well as in other hard-pressed communities, one of the greatest difficulties is that compensation amounts are low because of the poor quality and small size of existing agricultural plots, yet the cost of relocation is no less expensive than for people moving from more fertile areas. Corruption among local officials has been an ongoing issue, and in March 2001 three Yunyang farmers attempting to deliver petitions to officials protesting the illegal diversion of county resettlement funds to officials in Beijing were detained by police for months. They were later charged with disturbing the resettlement, leaking state secrets, and maintaining illicit relations with a foreign country, a reference to their contacts with the international press.

Despite aggressive attempts to put an end to it, various forms of corruption, from taking bribes to false resettlement schemes, are widespread and almost impossible for higher-level authorities to control. Both officials and migrants have devised a wide range of schemes to cheat the government and benefit themselves. High-level officials are in the best position to profit on a large scale, and there is a constant skimming off the top and inflation of project costs at lower levels. Overall, according to official Chinese figures, 463 million yuan (U.S. $56 million) in resettlement funds were illegally used for other purposes in 1998 alone. In 1999, the former director of the district construction bureau in Fengdu was sentenced to death for stealing 12 million yuan (U.S. $1.45 million), and the auditor general of the People's Republic of China estimated that 7.4 percent of the entire resettlement budget of 17.68 billion yuan (U.S. $2.14 billion) had been embezzled since work on the dam project began.

From local officials who falsify township records to show that people who have not moved or do not even exist have been relocated, to peasants who take their resettlement money and head for cities in the south, leaving behind their families in villages to be flooded, the number of scams using funds meant for long-term resettlement for short-term profits is enormous. Factories with government grants to create jobs for migrants have instead provided peasants with documents stating that they are employed and handed over cash settlements, taking a substantial cut for the service. This frees the enterprise to make a profit without the burden of operating a business or providing jobs or housing. The peasants end up with what seems like a large sum but no prospect of permanent employment. Some migrants succeed in investing in small businesses of their own and can do well with the sudden boomtown opportunities of communities in flux. Others, particularly those who have supposedly agreed to move to another town or county to the site of a fictitious job, may end up with no legal right to live where they are or any place else either. With their household registrations transferred to their nonexistent new homes, the relocatees become part of the floating population, without the right of residence or access to community services, but in the towns where they grew up.

Since 1949, reservoir and irrigation projects have resulted in the relocation of approximately 10.2 million people in China, a number that includes not only those who were moved at the time of construction, but now their children and grandchildren as well. Approximately two million people were displaced by projects along the Yangtze and its tributaries alone. Many of these large-scale water diversion projects, most of which were built during the Great Leap Forward and the Cultural Revolution, left millions living in poverty. In 1986, the Ministry of Water Conservancy and Electric Power began a 1.9 billion yuan program to assist five million people at forty-six reservoir relocation sites. In 1989 a Chinese government study estimated that seven million, or about 70 percent, of the reservoir relocatees, were still living in extreme poverty. The general economic improvement in China in the 1990s helped, but not enough. Li Boning, the head of relocation programs for the Three Gorges Dam, has estimated that of these *shuiku yimin*, or reservoir migrants, approximately a third are worse off than before they moved, a third have tolerable living conditions, and a third have an improved standard of living—not an impressive record for the past fifty years.

Previous reservoir relocations in China offer few models of success. Since regional planning for the Three Gorges Dam began in the 1980s, the government has emphasized the policy of developmental resettlement (*kaifaxing yimin*) in the hopes that by integrating rural migrants into an overall program of economic development, the results will be less economically and socially traumatic. In earlier reservoir relocations, migrants usually were given a one-time lump-sum compensation, but no assistance in adjusting to new farming conditions or finding employment. Once their compensation funds ran out, many were left destitute. The goal of developmental resettlement, much like the original Tennessee Valley Authority plan, is to use the dam as the centerpiece of an overall economic revitalization program, combining the creation of an infrastructure with modern roads and communication, outside investment, and the establishment of new townships and modern enterprises with improvements in education and training that will allow peasants to move into sideline industries and off-farm jobs. There has already been an enormous improvement in the standard of living in the region since the early 1990s, but how realistic the sweeping changes envisioned for the Three Gorges are, and over what time frame, remains unclear and in dispute.

Researchers from the Chinese Academy of Social Sciences have estimated that it would take about 370,650 acres—five times the amount of cultivated land to be submerged—to support the same number of people who were farming in the mid-1990s because so much less fertile land is available at higher elevations of the gorges. This shortfall increased after the 1998 floods when stricter rules prohibiting deforestation and the tilling of fields on grades steeper than twenty-five degrees were instituted or more rigorously enforced. Despite the original emphasis on moving near and higher, by the late 1990s it had become evident that it would not be possible to relocate so much of the rural population within the same region without causing serious hardship and social instability. In an attempt to reduce this strain, the government has increased the number of people moving farther away, both within the reservoir area and to the outskirts of Chongqing, and to other parts of China.

In recent years, government policy has been to give peasants somewhat more say in where they will go, but overall their input is still limited. When possible, villages are moved as a unit, whether to higher areas within the gorges or to more distant locations, to maintain social cohesion and provide community support in a new and sometimes unwelcoming environment. Individual farming families who have relatives in rural areas above the new water level of the reservoir may make their own arrangements and join them, but these migrants must remain in the countryside. Here, as elsewhere in rural China, legal migration to urban areas is rarely permitted, though many go nonetheless.

Thousands of peasant families, both from the countryside and the rural areas on the outskirts of the county seats, are being relocated to new townships near Chongqing, Wanxian, and other county seats in and above the gorges. Most are given a one-time compensation package, which is applied toward their housing in high-rise buildings. The families who move into these towns come from counties throughout the gorges and are largely self-selected from rural areas near existing towns rather than from villages deep in the countryside. They are the peasants with comparatively high levels of education, often middle-school graduates, who may have developed or want to develop nonagricultural skills or who are willing to try anything to give their children a chance to find work off the land. Like many of the peasants who have already left for the cities illegally, this group tends to represent those in the middle, neither the best off, who have reason to stay where they are, nor the worst off, who have neither the skills nor the resources to survive by anything but farming. The relocation bureaus are making some attempt to provide assistance in finding new jobs, but, for the most part, the new immigrants are left to themselves, and either they are exceptionally resourceful or officials are surprisingly cavalier about how they will manage. Some find positions in factories, more a possibility in Chongqing than in other cities, but many others find only temporary construction jobs or sell what they can on the streets.

As a foreigner visiting the Three Gorges area, I could not escape the notice of nervous authorities, and though official assistance was sometimes generously offered, it is difficult to obtain permission to visit anything but showcase sites. Local officials would prefer that any problems under their purview remain for them to solve or to keep to themselves, and that tales of discontent or mismanagement not reach a higher level or the general public, through the media or uncontrolled word of mouth. They fear being blamed for publicizing issues that everyone knows exist but are not faced openly, or worse, the discovery of problems for which they are in fact responsible. This anxiety frequently leads to a thinly disguised paranoia about anyone from the outside, whether they are from New York City or Beijing.

Almost all peasants forced to move complain that the amounts provided by the state are insufficient to start over and recoup their previous standards of living. Town residents may struggle to find funds for new housing, but the majority are employees who can expect their salaries to continue. In contrast, peasants are suddenly dependent upon their own ability to find new work or to adapt quickly to new agricultural conditions. Although many peasants would not hesitate to move into other jobs given the choice (in a survey conducted in Badong county in Hubei, for example, 87 percent said they would be willing to consider leaving the land),[13] the reality of life away from a familiar place where they and their families have always been known can be harsh. When I have asked officials how peasant families adjust, they are for the most part dismissive of any difficulties, saying that it makes no difference whether the new township dwellers come from Yunyang or Fengjie or Kaixian, because the places are all more or less the same. Officials repeatedly told me that the different towns have no special characteristics or differences, everyone is Han (the main ethnic group in China) and that after a short period of adjustment to city sanitation practices, there are no differences. This of course is not necessarily how the relocatees themselves look at it. In Yunyang, for example, 17 percent of the 358 people placed in factories in the new town or other communities during the first five years of the relocation program returned to their villages because they could not adjust to their new lives, and numerous other relocatees have found it impossible to become accustomed to indoor life and the fixed working hours expected of them in industrial jobs. Retraining rural residents in trades—teaching people to be tailors or mechanics—has had some limited success, but this only works if there are other people who need their services, and in many cases there are not. The result is that it will be necessary for the government to provide continuing stipends to people for whom neither suitable work nor agricultural land can be found, a number that may reach as high as 10 to 20 percent of the relocatees in some counties. This is a heavy burden for most local governments; as the welfare payments are currently structured (the payments are equal to the interest on the per capita sum designated for resettlement, which in most counties is in the range of six or seven thousand yuan), they provide only about thirty-five yuan (U.S. $4.23) a month for living expenses, not enough to live on anywhere.

The disruption is greatest for the more than 100,000 people who are moving to other parts of China. In the summer of 2001, 70,000 residents of the Three Gorges began an exodus to rural areas and new townships in eleven designated sites in Shandong, Shanghai, Hunan, Hubei, Jiangsu, Jiangxi, Zhejiang, Guangdong, Fujian, Sichuan, and distant parts of Chongqing municipality. Over 125,000 are expected to move to faraway locations by 2003. For peasants transferred to other parts of Sichuan or neighboring provinces such as Hunan and Hubei, where the language and terrain are similar, the transition will not be much harder than for other migrants settling in communities close-by. This will not be the case for those going farther away. Farmers heading off to the coastal regions of Shanghai and Fujian and other places where farming methods are entirely different and the local people speak completely unintelligible dialects may be there years before they feel at home. (Fujianese, for example, is no easier to understand for someone from Sichuan than German is for an American.) Experimental relocations to Xinjiang and Hainan provinces, in which many farmers in Wushan, Fengjie, and several towns in Zhongxian county participated, have been abandoned. There had been longstanding connections between communities in the gorges and both Hainan and Xinjiang (Zhongxian county had a joint agricultural development program with Hainan and many men from this part of Sichuan worked in construction in Xinjiang during the winters). Nonetheless, 90 percent of the migrants, although they went as volunteers, returned home within six months to a year. In Xinjiang, the migrants could not adapt long-term to the northern climate and barren land. In Hainan, the problems seemed to be mainly organizational, with relocatees finding land and housing conditions impossible and entirely different from what they thought they had been promised. Both provinces were dropped as relocation sites.

This is not the first great migration in the history of the Three Gorges region. In the early seventeenth century, during the years of famine and brutal killings at the end of the Ming dynasty, most of the surviving inhabitants fled. Faced with no one to till the land in one of the biggest and most fertile provinces of China, the new rulers both forced and encouraged peasants from Hubei, Hunan, Guangdong, Guizhou, Shaanxi, and other parts of China to leave their homes and take over abandoned fields and orchards. Many natives of the gorges and other parts of Sichuan trace their ancestry back to this great population shift known as *Huguang tian Sichuan*, or "Hubei and Hunan fill in Sichuan." New settlers were offered ownership of deserted lands and short-term tax exemptions, and stories still abound about peasants being dragged here in chain gangs to repopulate the depleted region. Even longer ago, during the Song dynasty, thousands of other people, descendants of the Ba and the ancestors of the Tujia who now live in the mountainous area near the Shennong River in Hubei, left the gorges for Fujian. Though the migrants have been indistinguishable from the people of Fujian for centuries, researchers have recently discovered elements of flute music in Fujian similar to that of the Ba region of the gorges. It is believed to be derived from the music of the migrants of a thousand years ago.

CHAPTER 11

FROM PAST TO PRESENT

IN MARCH 1997, EIGHT MONTHS BEFORE THE DIVERSION of the Yangtze, Chongqing was upgraded from a city to a municipality, permitting it to report directly to the central government. Although there was to be no Three Gorges province, the change in Chongqing's structure removed the entire Sichuan portion of the Three Gorges region, including Qutang and Wu Gorges, from the province and combined it with the city of Chongqing to give the new municipality an area about the same size as Austria. Thirty million people, roughly the same population as Canada, live in what is now Chongqing municipality. The region has always lagged behind China's coastal areas, but since 1997, once construction of the dam became a certainty, an infusion of investment and new streamlined economic policies had an immediate impact on the area. A shift in central government policy and resources to develop western China was launched in 2000, bringing a variety of preferential taxes and other benefits that also encouraged investment. New funding and new officials have poured into Chongqing, resulting in rapid improvement in conditions and opportunities for some, but not for all, a phenomenon accompanied by a sudden swell of corruption as people have attempted to exploit the new pots of money. Despite this, and the almost wholesale destruction of the old city, there is little nostalgia for the past. Unlike Beijing, where a once-beautiful city was destroyed during the fifty years after 1949, Chongqing was always an ugly, uncomfortable place. For the most part, it has been improving lately.

For months on end the heat in Chongqing wraps around you like a wet, dirty towel, and for most of the rest of the year you cannot see through the mist. The old World War II airport where I once spent so many days waiting for planes that never came because of the endless fog has been replaced by a new one on a different side of the city. Instead of a careening ride over a mountaintop, you now take a dangerously fast trip along a four-lane highway lined with what are called "Spanish villas" here. These huge stucco houses, with red-tiled roofs that look like they belong in southern California, and cost almost as much as if they were, are owned mainly by the entrepreneurs and deal makers of Chongqing who have found a way to profit from the opening of the city to foreign trade and the massive boom in building and services. In the city center, even with the opening of several new bridges across the Yangtze, the streets are blocked by endless, crawling traffic jams. A multitude of skyscrapers with revolving tops have appeared downtown, replacing seven-story walk-ups and the dark and winding alleys. Signs informing the residents that their homes will be destroyed in a few weeks to make way for new buildings are regularly posted on the walls of old neighborhoods, and their inhabitants packed off to high-rise blocks, often to wait years in temporary housing. From the bamboo shacks of people whose homes have been razed, to the glitz of the international hotels, new construction is all around.

Some landmarks still stand. The Renmin Binguan, the People's Hotel, one of hundreds of hotels in China with the same name, is a huge and imposing structure with a large round auditorium resembling, from the outside, the Temple of Heaven in Beijing. Despite its perpetually leaking walls and ceilings and the oversized rats, in the late 1970s and 1980s, the Renmin was the hotel of choice for Chinese cadres, foreign tour groups, and business travelers. In those days, the hotel had a pegboard where people were supposed to hang their keys when they went out for the day. Since no one at the front desk made an effort to distinguish the guests from one another, everyone in the hotel had complete access to all rooms, which resulted in a constant state of petty thievery and confusion about who was staying where. The lobby was always filled with distressed guests of all nationalities, many of whom found it impossible to make themselves understood in any language at all to the surly young men behind the reception counter who were the sole link to food, rooms, and transportation out. Regardless of what reservations had been made or how long ago, there were never enough rooms. Check-in was an unending negotiation for everyone. When there really was no space, people were billeted in the staff dormitories or on mattresses in the dining room. Most often, rooms went to the customers with the most stamina, the ones who had perfected the right balance of groveling, apologizing, threatening, and pulling strings.

The hotel has been remodeled, and thousands of people now gather at night on its huge grounds for ballroom dancing and concerts. Most of the old buildings around it have been demolished and replaced by apartment complexes with fountains and swimming pools for the wealthy and high ranking. Government funding for a Three Gorges museum was approved in 2001, to be built across from the hotel once the land is cleared. Officials say that the museum will provide a comprehensive overview of the culture of the Three Gorges region, housing artifacts unearthed during preparations for the dam, for which no suitable location has been found locally. The decision to move some of these objects to Chongqing has been an occasional source of tension between local and regional authorities, as many communities being flooded would prefer that archeological and historical finds remain nearby.

Despite new express buses and more frequent airline service, the harbor in Chongqing remains the hub of transportation for almost everyone going anywhere in this part of China. On most days it seems like they got here all at once. Since the beginning, Chongqing's life has been built around the river. Millions of travelers get on or off a boat on these docks every year. The Chongqing Shipping Company alone employs 18,000 people located here and elsewhere in the municipality. The harbor also attracts an army of porters with poles waiting for arriving ships. Mostly men from the Three Gorges and the outskirts of Chongqing, along with a contingent of unemployed and marginally educated city residents, they stand by to carry anything that must go up the three hundred steps to the road. The men haul crates, produce, suitcases, people on stretchers bound for city hospitals, overweight tourists, and anything else that cannot make it up to the top on its own.

All harbors have their own distinct sounds, smells, and rhythms immediately identifiable to those who know them. In Chongqing, the harbor whirls with the motion of a thousand ships set loose and careening toward one another from different directions. Instead of the orderly conversations between boats along the rest of the river, with their clear long and short horn blasts, here whistles and bells and roars of every pitch and volume compete with loud voices distorted by bullhorns. Men shout out demands for docking space in Sichuanese and Mandarin as they challenge other ships for space. The water and the air give off a scent of raw sewage, diesel fumes, sweat, dust, gasoline, and cooking oil mixed with traces of a sea-like freshness in the breeze.

The pace of transportation has gotten faster. More people are going places. The old East Is Red passenger and freight ships (which sail on, though under the less romantic names of Chuan, Jianghan, and Jiangbei) could take as long as two weeks for the voyage from Shanghai to Chongqing, stopping once or twice daily at major and minor ports. Now express ships and Yugoslav hovercraft, purchased soon after that country's dissolution and now managed out of the monumental shipping companies in Wuhan and Chongqing, tear up and down the river. These have transformed a twenty-four hour voyage between cities in the gorges into an afternoon's trip for those who can afford the tickets. On the speeding hovercraft, hundreds of passengers sit in glassy-eyed silence, transfixed by Arnold Schwarzenegger and Hong Kong martial arts films dubbed into Mandarin Chinese. The soundtrack is loud enough to wake the dead, but not the few passengers who have managed to fall asleep. Purple velvet curtains give the cabin a peculiar pinkish glow and shut out not only the glaring sun but all traces of the view.

On these secondhand hovercraft, as in the towns themselves, the sense of modernity is deceptive and one can suddenly come face-to-face with the reality of how far away river life is from where one might want to be. During a trip from the county seat of Zhongxian, I watched an injured man being brought aboard, immobile on a stretcher. He was accompanied by his parents, wife, three brothers, and a doctor and was headed for a hospital in Chongqing, where we were supposed to arrive in two hours. Near Fengdu, the boat stalled in a torrential storm and we were towed to shore. The sick young man, soaked from the steamy rain, was carried off and put in a shelter on the dock. Four hours later, back on the hovercraft, he began to fail. His doctor, probably a village medic, began artificial respiration while his patient faded in and out of consciousness. After a while another passenger, a more highly trained doctor, took over, but to no avail. While Hong Kong disco music exploded through the cabin and rain blew through the door and sloshed around the stretcher, the patient died. The patient's mother cried out and the people on the boat looked uneasy and uncomfortable. It was bad luck to have a death on board, someone murmured, worse luck for the young man to have stopped in Fengdu, the City of Ghosts. In a place with so many spirits of the recently dead wandering about, ready to steal other people's souls, he was an easy target in his weakened condition.

The old junks with their stately double sails unfurled against the wind have almost disappeared. Too expensive and not practical, explained a boatbuilder in Wushan. His grandfather knew how to build them, maybe his father too, but only a wealthy family could afford one, and what would they do with it nowadays? His advice: "An old sailboat can't go as fast as a motorboat. If you want to make money, get together with friends and buy a motorboat big enough for at least twenty people, and run a ferry. That's where the business is." Despite his recommendations, many families stick to their old home-built sampans. In the early dawn, the flat three-board boats still drift across the Yangtze, often with just one man aboard, hunched over with his fishing pole or net on the edge of the narrow raft. Occasionally men are out with their cormorants, birds traditionally used to capture fish. With a ring around its neck to keep it from swallowing its catch and tethered to a string to make sure it comes back, the bird sweeps into the river, plucks a fish from the water, and returns to the boat, where it is rewarded with a fish it can swallow.

An anthology of stories from Shibao Block includes one about a man selling tofu on the main street around 1983. A foreigner, a Westerner, stops to buy some, eats it, and goes away. The vendor is so aghast at seeing someone from another country in Shibao Block for the first time in decades that he does not know what to do. He mulls over an undescribed but clearly unpleasant incident between his father and a Western official in the town before the founding of the People's Republic in 1949. Deciding to take a stand against this unexpected incursion, he throws away the money. This too upsets him, and as he sits there brooding, a cadre comes by and tells him to think it over. The Westerner paid him good money for the tofu, and there may be more coming. Foreigners have money to spend. The story ends there, without any indication of whether the tofu man will find the crumpled bills he has tossed away. He is still too shocked that people who have been evil for so long can now walk the streets and buy food like anyone else.

Anyone who has visited China in recent years knows that the tofu man did indeed take the cadre's words to heart, and that he now charges all foreigners and out-of town-Chinese at least double the going rate for this local delicacy. My first trips along the Yangtze were made about the time that this beancurd maker was wondering what to do with his money. In those early days of tourism, I led a parade of foreigners from port to port, keeping them happy with bottles of Johnny Walker Black Label scotch while trying to maintain the equilibrium and good nature of their Chinese escorts with apologies, Japanese alarm clocks, and signed affidavits of responsibility for everything that went wrong. It was a time when much was new, or when the old was coming back in a different form. All along the river people were suddenly doing things they had not done in decades. There were things to sell and buy, new businesses to start, places to go, and strange travelers from faraway places. The river was a place of romance and intrigue, real and imagined, for Chinese and foreigners alike.

For older visitors, the romance was from the past, an inevitably disappointing search for traces of the life they remembered before World War II. Old cadres appeared on the boats without explanation, possibly inspecting

Figure 11.1 Sign in Fengdu from the summer of 2001 showing the daily countdown (704 days) until the city is flooded.

the captain or the passengers, or simply taking the easiest route home. They would stare silently at the passing mountains and break into occasional tirades as we passed boats filled with Japanese tourists sailing through a region once terrorized by their countrymen. Elderly Americans and British missionary and merchant children who had grown up in Sichuan hill stations or foreign concessions often came back for the first time since the 1920s. They grew quieter as the weeks wore on and they discovered the almost total obliteration of everything they had known. Men who had sailed on the gunboats in the 1930s marveled how the modern cities of fifty years ago had become desolate and dreary.

The ship captains were men who had navigated the river for thirty or forty years, who had come to know every rock and shoal in the decades since they first shoveled coal or boiled water on a boat along the Yangtze. On tourist ships, the captain and the ubiquitous political commissar were party members who had the political and personal stamina to deal with dozens of foreigners, homesick interpreters from the city, sixteen-year-old crew members never before away from their villages, special police, and whoever else showed up. The crew quarters ran like a school dormitory, supervised by good-natured but often exasperated pursers and boiler chiefs responsible for making sure that nothing untoward happened. It was up to them to see that romance and homesickness, bad tempers and laziness, were kept under control and out of sight during the months of close confinement in the crowded lower decks.

At this time, almost anyone under forty working for CITS, China International Travel Service, the government organization that handled most arrangements for foreigners, had just begun his or her first job after a long delayed college education following years in the countryside or in a factory. These young men and women staffing the boats and explaining Chinese history in Japanese minibuses were among the handful of students who had passed the university entrance examinations after they were reinstated in 1977. Students returning from the countryside had to compete against the tens of thousands of others who also waited up to a decade for this opportunity for an education. The hard-won places in recently reopened or reorganized colleges and universities offered a largely normal academic curriculum, rather than a completely political one, for the first time since the beginning of the Cultural Revolution in the mid-1960s. For these interpreters and guides, mostly the sons and daughters of academics and high-ranking cadres who had expected to follow in their parents' footsteps, shipboard life was easy, but translating stories about rocks and watching kindergartners sing ad nauseam was not what they wanted for a lifetime. In those first years of revitalized tourism, work assignments, like their earlier sentences in the countryside, were supposed to be forever, and it still took years of negotiation, flattery, and bribery to change jobs.

The English-speaking employees on the ship, unlike engine room apprentices and waitresses, worked in close contact with the foreigners on board and often talked about their lives with the Western passengers and staff. The opportunity to converse with someone from the outside, with an entirely different perspective on life, sometimes provided an outlet for the frustrations of the educated young people, but it also made it all the more easy for them to run afoul of the authorities. There was always someone watching and monitoring. Each ship not only had the political commissar, but an assortment of other people to keep track of what everyone was doing. This included a number of Communist Youth League members and public security or foreign affairs officials who, depending on their goals and personalities, could haphazardly but vengefully scrutinize all interaction, sometimes causing enormous difficulties. The sense of isolation and semi-freedom on the boats led to wistful crushes and complicated rendezvous, with small boats paddling from cruise ship to cruise ship, and propositions and promises and misunderstandings. Intense relationships were negotiated while sorting out squabbles over bridge between elderly ladies or drinking late into the night with bored and sleepless businessmen whose presence made it possible for two others to be together in public. Real privacy came at great risk, but it was one often taken.

On the luxury ships, the river swirled slowly by behind floor-to-ceiling glass windows. Passengers could watch a panorama of sights and sounds and smells, enter into this environment briefly, and then withdraw. Local people did the same in reverse, though they retreated to a harsher world. It was not unusual for guests who forgot to draw their curtains to awaken in the morning and find a small fishing or vegetable boat docked against the window, the faces of the men pressed against the glass as they discussed in the loud and lilting cadence of the Sichuan dialect who and what was in the room and what they thought of it. Though a group of foreigners could attract a crowd of hundreds, and local people might assume travelers from Beijing were Japanese, most initial interactions between foreigners and local Chinese were simple and straightforward. Familiar with missionaries from the early part of the century, elderly women would come up and converse with me in Chinese, taking for granted that I would be able to respond. Sometimes they would ask questions about Europe or the American Midwest as if there had not been a gap of fifty years since their last such conversation. Others, educated in the mission schools once scattered throughout the region, approached us in resurrected childhood French and English.

For more sustained contact, no one quite knew what the appropriate lines were between Chinese and foreigners, or even between local people and Chinese from other parts of the country, or what to do if they were crossed. The guiding rule, enforced by each work-unit, was that no Chinese who came into contact with foreigners through his or her work was allowed to have a personal relationship or friendship with them. How this was interpreted could cause all sorts of problems, as happened when a young interpreter on the ship, Miss Chen, had the misfortune to be on board when an elderly passenger died, a not infrequent occurrence given the average age of foreign tourists at that time. After the death, Miss Chen assisted the widow with the paperwork, and arranged to have the body cremated. When the bereaved woman decided to leave the ashes behind to be scattered in the Yangtze (her husband had enjoyed his trip so much that she wanted him to stay), Miss Chen helped take care of that as well.

A few weeks later, the widow wrote a thank you note. She added that her house in California was big and empty and she would like to invite Miss Chen to study in the United States. The girl declined, but soon thereafter the political commissar informed her that he had learned that she had attempted to establish and use personal connections with a foreigner in order to leave China. Unbeknownst to her, the widow had kept writing, but the incriminating letters were intercepted by company censors and translated with an unfavorable edge, and sent on to a higher level. The result was that Miss Chen spent the next few months writing weekly explanations of why someone would offer to help her go abroad to study. She cited the outgoing nature of Americans and the loneliness of a recent widow, while I called California to stop the flow of letters. In the end, her leaders concluded that she was not necessarily at fault, but if she had done nothing wrong, why would anyone have suspected a problem?

Most officials these days have moved on from worrying about who else might be benefiting from foreign contacts and are busy establishing their own. Though all ships still have their political commissars, they are rarely the imposing and sometimes feared figures who not long ago made sure that the crew understood the spiritual pollution campaign and Deng Xiaoping Thought or decided that *The Exorcist* was not an appropriate movie for onboard viewing. On some of the tourist ships, though they may still oversee an occasional Communist Youth League meeting or run awards ceremonies, it is not uncommon for the political commissars to direct the ship's brass band or coordinate rehearsals for the weekly fashion show.

There are now over fifty ships designated for tourists on the river, some massive in size. Built in the shape of dragons and pagodas and great mirrored fortresses, they speed from one scenic spot to another, keeping their passengers entertained with in-cabin videos of the places they have seen or will see, karaoke, and nightclub acts. Barely a generation apart, the lives of the young people who work on the boats have little in common with the experience of their predecessors of fifteen years ago. Gone for the guides are the once *de rigueur* presentations of class background and how it affected their lives. In the old days, not so long ago, passengers would gather around in the overstuffed armchairs on the forward deck while Mr. Yu explained that his father was a poor peasant and his mother had been classified as a lower-middle peasant, and how their lives had improved because they had fought with the Communists from early on. Then, Mr. Li, assigned to the ship for a week by his office in Beijing, would more reluctantly tell the story of how his grandfather had been a famous scientist who had studied in America, his parents professors who had been persecuted during the 1950s and 1960s, and how his father had died in a labor camp, and his mother was now teaching French literature again. Such stories of tragedy and victory, of being done in by the landlords or driven out by the peasants, used to be common in a society lurching from one cataclysmic upheaval to another. For young people, this is now part of their parents' past. Workers in their late teens or early twenties will occasionally mention that one of their parents was able to get an education, the other not, because of the Cultural Revolution. Others nod, knowing generally why, if not the details. For many young workers and students, their relatives' pasts are shadowy, for though families vary, there are many middle-aged parents who say little to their children about either painful times or adolescent fervor they would rather put behind them.

On the tourist ships, the staff on the boats now have Western names like Peter and Cameron and Dahlia and Chairman, and they call Chongqing and Shanghai on cell phones, and stop at the Internet café in Wushan. In contrast with the past, when the staff vacuumed or translated and then had plenty of time left over to talk to the passengers or to play cards and sleep, the young people on the boats today have a more rigorous existence. Recruited by the big cruise lines and tourist companies through schools and newspapers and friends, they not only have to be able to fill thermoses or maneuver a tray around the dining room, but also must sing and dance, play an instrument, or have some other demonstrable talent in addition to their regular duties. On some ships, like the Victoria Cruise line, the girls from the kitchen and dining room perform complicated folk dances and keep up an exhausting schedule of rehearsals, costume fittings, and choreography, while the men from the boiler room play traditional stringed instruments like the *erhu* and *pipa* and the floor boys sing excerpts from classical Chinese opera and break-dance. After the performances are done, the young staff throw themselves into the macarena and dances new along the river, like the Bunny Hop. "Kick me, shove me, love me, do the Bunny Hop," go the lyrics to the tape, "Make me, break me, take me, hop, hop, hop."

No longer assigned by the state to their jobs, the young employees are thoughtful about different things than their parents were and more occupied by the future than the past. It is easy for recent graduates today to speak out loud about what they might like to do, to talk of spending a year or two on the boats to get experience and make contacts, and to plan to move on. A cashier on one of the ships, a twenty-two-year-old girl named Xiao Ma, described her tentative plans for the future, based on her mother's counsel. Xiao Ma explained that her mother, a businesswoman, understands her very well and knows how the world works and how it has changed since she was young. She has advised her daughter to continue her work on boats with foreign passengers for a few years to perfect her English, learn about the outside world, make contacts, and improve her skills. When she is about twenty-five, she should look for someone to marry, for by then she will have the judgment to make a good choice. In another year or two it will be time to have a child, while she is young and energetic, and to think about finding a job with a Chinese or foreign company involved in international business. By the time she is thirty, she will be in a position to advance rapidly in whatever career she has chosen. It is sensible and practical advice, both in terms of Chinese society and what Xiao Ma thinks she wants. She is ready to follow this path, but unlike her mother's generation, she does not feel constrained by outside forces. She does not necessarily expect the future to be easy, but she expects that if she makes the right moves and choices, she will have many options.

In the Three Gorges area, there is still a chasm between the lives of the people who live there and those who float by, and despite the decades of economic reform, the gap between the towns and the countryside is equally great, both financially and culturally. According to a survey of residents of Badong county in 1992, only 72 percent had heard of the dam and only 28 percent knew where it was to be located.[1] A few years later, teenage students in Shibao Block were recounting garbled stories of being recruited to build bomb shelters in Yichang, when in fact the big hole they were to dig was the new dam. There are now notices everywhere about the stages of flooding and relocation, and the number of days until the city residents will move, but their meaning does not always seem to register. Peasants and city residents alike seem to have little grasp of what is to come next beyond what will happen to them and their neighbors. Knowledge is particularly limited for people migrating to other parts of China, whose total information is based on the brief visit of a few village officials who have been sent out to far away relocation sites and shown the best of what might be, but often is not, in order to convince their neighbors that it will be a fine place to live.

Even in Wanxian, where most of the city is to be moved and there are signs everywhere indicating the dates that this will happen, exactly what this means is often lost on the city's residents. There are 175-meter signs, marking what will be the new water level, outside the entrance to West Hill Park and in front of the 1930 clock tower just inside the gates of the park. It is a sprawling, much-used park, with children's amusement rides and long shaded pathways leading through groves of semitropical plants. The clock tower, once the highest structure in Wanxian, used to be open daily, and for a few cents you could climb to the top for an unequalled view of the harbor and river. It is closed now because the number of people buying tickets no longer justifies an attendant's pay. A woman selling film in front of it says that few tourists come to Wanxian these days, and although local people went up the tower over and over again when there was nothing else to do, now that there are video stores and movies and dance halls and karaoke, why bother with more than one trip up? Passersby told me how lucky the city was that the water would reach only to the gates of the park, and that one of the best places in Wanxian would be spared. In reality, though the water will come just to the base of the old clock tower, the harbor will not be lapping at the steps of the park entrance as imagined. The tower, a landmark in Wanxian for over seventy years, and everything else at this level will be cleared away for the new docks and a wide modern road connecting what had been the old city to the new.

The sloping hills outside of Shibao Block and the dirt roads that led out to the firecracker factory through high fields of sorghum are now occupied by concrete two- and three-story buildings, but the old town is still active. As long as boats still dock and tourists still come to see the high red pagoda, the trinket sellers and restaurateurs will hang on to the charming old town for as long as possible, regardless of what they think of the advantages of one place over the other. Here, as in many other places along the river, the old and new towns exist simultaneously during this strange interlude when past and future overlap. Some old houses have been dismantled by their owners in order to use the materials for their new homes. Other families have two houses for now, a practical solution to problems ranging from running a shop in the old town to keeping a child near a school that is yet to be relocated or taking care of an elderly relative who wants to stay where she is.

Figure 11.2 Window display in Shibao Block

In Shibao Block as elsewhere, parents continue to worry about how their children will earn a living and have little faith that the dam is going to solve problems like the lack of jobs that has led to the town emptying itself of young people. Since the groundbreaking in Sandouping in 1994, up to 30,000 people have been working around the clock on the dam. Engineers, demolition experts, electricians, and other specialists have relocated to the site for a ten-year stay, often lured by high pay. Laborers from rural communities along the river can also make as much as three times what they might earn at home shoveling coal or working on local construction sites. For a seven-day week, they earn 2,000 yuan (about U.S. $250) in a month, a rare opportunity to save for a new house, business, or marriage. Still, some of the men are skeptical about what comes next and where it will leave them. "It's all short-term work," said one farmer in Zhongxian county whose son is a laborer on the dam, reflecting the views of many who have family members there. "Our sons can go to the site for a few years and dig holes, but what does it teach them? When they come back, there will still be no work, whether we're here or higher up on some other mountain."

On the streets of the river towns, conversations quickly turn to the corruption of local officials. In Shibao Block, perhaps because of its small size and the relatively egalitarian layout of the new town, the focus still seems to be mainly on the implementation of family planning rules rather than the consequences of the dam. In contrast to many parts of rural and small-town China, where there is a clear superabundance of little boys, there seem to be more girls in Shibao Block. This is apparently the result of sheer bad luck. The town walls and bulletin boards are plastered with messages about family planning, from the usual "It is glorious to have only one child!" to equally ignored phrases like "A happy family has only three members!"

Despite such efforts, the town seems to have come to terms with two-to-three-child families, a situation that is common throughout rural areas of China when the firstborn is a girl. This allows parents a few chances to have a son, and failing that a couple of daughters whose combined strength or wit, or years in a southern factory, will be of some use until they marry. It also gives the town a considerable source of revenue from these illegal children. Opinions differ about the official amount of the fines, reportedly set "according to family circumstances," as well as where the money goes. Babies are routinely introduced as "my ten-thousand-yuan child," not because the fine was in fact ten thousand yuan, which would be about five years' income for fairly prosperous peasants in this region, but because the term indicates such a huge sum of money, as in "my million-dollar baby." Almost everyone claims that the fines are far higher than what the government actually permits and that the difference goes to the collectors. People here seem to accept this with a kind of aggrieved stoicism, viewing it as part of an age-old ritual that brings a certain kind of equilibrium. No one much wants more than two or three children anyway, and for a price, the authorities are happy not to interfere.

In other towns, complaints are more often about the mismanagement or embezzlement of relocation funds by local officials. All one has to do is sit quietly in a public place to hear the frequent tirades about the government misuse of money for personal gain. In a teahouse on the dock in Fengjie one evening, I noticed an older man on a rattan couch reading his newspaper. Though he did not speak to me, he occasionally told the waitress to get me fresh hot water for my tea with the authority of a person of some importance. A few other men joined him, fanning themselves on the hot summer night. As I watched the comings and goings on the dock, I caught fragments of what soon became a noisy discussion about corrupt cadres, half the building funds disappearing, and if Jiang Zemin thinks this is a good thing, then let Jiang Zemin come live here. My semidisguised attention was eventually noted, and the older man asked if I could understand them. "More or less," I answered, "for you were speaking loudly." He said, "These men are not too happy about corruption in China. They think it is a big problem," his tone suggesting he did not agree or was at least weary of the issue. I told him that I, too, had heard it was a problem, though not just in China. Apparently satisfied that I had reaffirmed his point, whatever it was, he repeated this. "You see, she says that corruption is a problem everywhere." One of his companions spoke up, "And he's the king of corruption, our party leader here." The party leader, or whoever he was, ignored this comment and changed the subject to how the American president Little Bush (as opposed to Old Bush, his father) had been stirring up trouble and ought to talk to his father more. The pitch of the conversation became tense again, and it was clear that my participation was not welcome, so I returned to my study of the river. I heard the men ask how does this—whatever this was—the move, the dam, economic reform—benefit the ordinary people, and then the annoyed response of the senior cadre that there was a foreigner sitting right there, and you do not need to say such things in front of a foreigner.

What is said about official corruption here is strangely similar in content, though not in context, to the conversations one heard in Beijing in the spring of 1989 in the wave of dissatisfaction and ebullience that preceded the Tiananmen crackdown, but without any sense of idealism that this might change. There is little sign in the cities of an energy that might lead anyone to take to the streets. Instead, there is a constant low-level grumbling, tolerated as long as it is kept at a local level and no attempts are made to get attention elsewhere, and offset by the prospect of new housing and the hope of more jobs. For those who bring their complaints to Beijing or the attention of the press, the consequences of real protest can be great, as shown by the arrests of petitioning farmers in Yunyang in 2001.[2] For most people, the risk and bother exceed any possible gain. What is the benefit of complaining too loudly about officials who will only make your life more difficult in the end? Furthermore, everyone here has become so used to the corruption around them that they no longer see it, say some people in explanation. As a Chinese proverb puts it, *shi er bu jian, ting er bu wen:* you don't see or hear what is familiar.

CHAPTER 12

THE EVE OF THE FLOOD

IN A BOOK ON THE THREE GORGES DAM put out by the State Science and Technology Commission, the authors put the new dam in the context of other great projects in China's history:

> The country has witnessed the birth of generations of outstanding talent that have distinguished themselves for the centuries. Looking back, our ancestors have created countless miracles. These include valuable and extensive experience in harnessing floods from the era of the Great Emperor Yu, the splendid cultural heritage of the Confucian era, the magnificent Great Wall, the beautiful Summer Palace, and the Gezhouba Dam, the first modern wonder in the development of the Yangtze from the Mao Zedong era. We now have an epic undertaking—the Three Gorges project—from Deng Xiaoping's era. The project will not only make a contribution to the present but will also benefit the nation and people for centuries to come. The future of the Three Gorges will be more beautiful than ever.[1]

The Grand Canal, built by the emperor Sui Yangdi (569–618) in the early seventh century, exemplifies another of these miraculous development projects. A ninth-century poem by Pi Rixiu (ca. 834–883) praises the canal as an achievement almost comparable to the ancient Emperor Yu's successes in flood control.

> People say that the Sui dynasty collapsed because of this canal
> But even today we still depend upon it for our transportation.

If the Sui Emperor had not used it for his personal pleasure
We could rank this canal as great as Yu's flood control project.[2]

In Chinese history, the rulers who were most successful in harnessing the resources to create the great infrastructure projects for which they are remembered were often considered tyrants while they were alive; not surprising, considering what it took to complete such works. Today's construction methods are less brutal, but contemporary leaders still brook no serious opposition to their plans. Despite its costs, taming the environment is almost always seen as good in China, and it is understood by almost everyone there, if not accepted equally happily, that completion of the Three Gorges Dam requires making sacrifices. As one former official from Fengjie explained, the dam is creating both a burden and an opportunity, a burden because of the expense and inconvenience, an opportunity because there is the chance that it will lead to new jobs, better education, and something approaching economic and cultural parity with the rest of China. Many environmentalists and engineers have long advised that a series of smaller dams on the tributaries of the Yangtze could achieve equally successful results both in hydroelectric generation and flood control, with far fewer harmful consequences. People who live here know that while small dams and reservoirs may cause local improvements, or local harm, they do not bring the wholesale commitment to change that comes with a project of crucial importance to the image of the nation itself. After forty years of minimal government attention to the Three Gorges, except as a site where a dam might one day be built for the benefit of the rest of China, local city dwellers, and some peasants as well, feel that it is time for the government to do something. For urban residents, the chance of a positive change is worth the disruption, and if plans to develop and revitalize the area fail, then at least it will be the government's fault, a release from local responsibility for problems caused by geography or unemployment.

Information about the possible consequences of the dam and international opposition to it is now relatively easily accessible through the Internet and generates some interest. Not unexpectedly in China, channels through which one can voice opinions anonymously are the most likely to elicit expressions of concern about the dam or anger over government policy. A call-in Voice of America television program about the Three Gorges Dam, broadcast in China in July 2001, drew numerous comments and questions about the dam, what kind of environmental damage it might cause, and whether there was reason to fear that it would not work. Chinese language Internet sites and chat rooms focusing on political and environmental issues offer other venues for exchanging ideas or venting anger about the dam project and the officials involved, though mainland Chinese say that they believe that many of the participants are overseas Chinese or citizens of the P.R.C. living abroad. A typical comment comes from a contributor who facetiously signed himself Zhang Guangdou, the name of one of the principal designers of the Three Gorges Dam. He wrote, "In regard to the construction in the upper section of the Three Gorges, people say that our scientific research departments made calculations and simulations . . . that supposedly proved that sediment would not accumulate in the Three Gorges. They are lying to themselves. One can only conclude that the scientists are changing the variables of the equation to satisfy the goal. In the future when everything is clogged with sand, you will see how trustworthy these hydrologists were. You must remember that in 1958 during the Great Leap Forward there were many experts who believed that the way to make steel was in the backyard. . . . The result of using the construction of the Three Gorges Dam to build a monument and make a place in history will ultimately reveal the proponents of the dam as criminals forever."[3]

In Beijing, engineer and journalist Dai Qing and a few others continue their fight against the dam, but within China the issue no longer arouses the kind of concern that her campaign did in the late 1980s and 1990s. In Chongqing, the only city in the upper Yangtze where there are real centers of research or academic activity, universities have a variety of projects concerned with the development of the area, but there has been no regional surge in environmental or other sorts of activism stimulated by the Three Gorges Dam. Government officials and others opposed the dam for many years because of the dangers of sedimentation and the heavy burdens it would place upon the city, but for the most part, both regional authorities and intellectuals have turned their energies elsewhere. A Chongqing Shipping Company employee turned successful businessman, part of a new group of younger educated entrepreneurs with international access and knowledge of outside opinions, told me he is not sure why so few people find the dam of interest. In his opinion, no one in Chongqing would ever think about it or know it existed if it were not for television, probably because Chinese people rarely are concerned about issues unless they are related to their personal well-being, mainly whether or not there is enough to eat. He pointed out that this was at the root of all the peasant rebellions. He spent years working as a river guide on the Yangtze and has read Dai Qing's books laying out the potential negative consequences of the dam. Sympathetic to the environmental and cultural issues, and knowledgeable about what may happen if things go wrong, he takes a practical approach, like most people here. Personally, he is more interested in the new business opportunities that have emerged in the city in the past few years, a time period that coincides with the building of the dam, and he doesn't worry about what might happen in four or five years. There is plenty of time to make a profit and worry about other things later, for him and other motivated entrepreneurs in Chongqing.

Foreigners constantly seek out Dai Qing to ask her how to stop the Three Gorges Dam. She tells them to keep after the foreign companies to stop their investment, particularly the European firms that have the greatest financial interests. I have asked her, with the dam well on its way to completion in Sandouping, whether she really thinks that there is any chance that the government will stop it. She said that she did not know, but one must try, and that perhaps the result would be a lower dam or some other change with fewer environmental consequences. Dai Qing's sustained energy and tenacity against this project shows no sign of abating, but as she says, she is one of the youngest opponents of the dam left. Huang Wanli, a long-time opponent of the project, died in 2001 at the age of ninety. Li Rui and the handful of surviving men who have spoken out against the Three Gorges Dam for decades are now also in their eighties and nineties and have not been replaced by a new generation. This is in part, says Dai Qing, because senior members of the government equate opposition to the dam to defiance of the government, a risk that few younger people can afford to take. Some environmental groups, such as Friends of Nature, headed by Liang Congjie, have expressed concerns about sedimentation, pollution, and the relocation, but the dam is not their main focus, nor is it their intention to become or to be seen as an opposition force.

Chinese intellectuals, mainly based in Beijing, periodically write letters to President Jiang or the State Council about the environmental impact or cultural losses resulting from the dam and recommend a reconsideration of plans. They have the freedom to do this, though their efforts are rarely acknowledged or acted upon. In a white paper sent to the government in the year 2000, one economist advocated that rather than attempt to develop infrastructure and industry in the Three Gorges region, the reservoir area should be turned into an environmentally protected zone to be used for camping, hiking, and tourism. The transformation into a giant park would be achieved by reducing the population of the cities by two-thirds, a size that could be sustained by basic agriculture and newly created service industries for visitors. To do this would require the relocation of even more people than are currently being moved, but, as the author of the paper said, China has great experience at this. Other people are more cynical, or perhaps more realistic, about what the future holds. As another former shipping company employee put it, it is the people who may end up hungry who will make themselves heard. They have the drive to fight threats to their livelihood and are the ones the government has to worry about. Public protests and other attempts by citizens to bring attention to harmful policies or the misdeeds of officials have the potential of awakening the sudden and never-completely-predictable anger that can erupt in China. Government leaders, from the local magistrate to President Jiang Zemin, fear this most.

⁂

However long it takes for the river to rise to its final form, the Three Gorges have become a different place already, one that will continue its evolution for decades to come. After the initial increase to 443 feet above sea level in 2003, the water is due to rise to 512 feet in 2006, and then to reach the full reservoir level of 574 feet in 2009. Some officials and academics familiar with the project predict that the final increase will be put off beyond 2009 to ensure that any technical and environmental problems are investigated and dealt with early on, but this is unlikely to be determined or publicly announced for some years. Most of the new towns will be completed before 2003, and the lower levels of all the old towns completely evacuated. Some communities are already gone.

Zigui, home of the Qu Yuan temple, was the first town to be completely relocated when, in 1999, the residents moved downstream to Maoping, not far from the dam site. The old city remains where it was, crumbling and abandoned, an odd specter of what is to become of all the other towns along the river. Always a favored spot to enjoy *huoguo*, the spicy Sichuanese hotpot, the harbor road was once a busy street. Now all that is left there are a few concrete shells of what had been shops and the three *huoguo* restaurants. These remain open, catering to the ships docked below to avoid nighttime navigation in the gorges. In the late evenings, the smell of boiling oil lingers in the street, but the only sounds are the shouts of drunken men enjoying a night in port with no ship to steer or boiler to tend.

Beyond the restaurants, where most of the town stood, the road led into the darkness past deserted housing, a ghost city inhabited only by a few caretakers and drivers who provided transportation for Qu Yuan's visitors and for officials in charge of the relocation and removal of the city. No one seemed to know exactly how or when the structures would be taken down to clear the river for navigation, but in Zigui people envisioned that what had been started with pickaxes would be completed with a giant chain of explosions up and down the Yangtze. Exactly this took place in March 2002, when what remained of Zigui was blown to bits. A slow process of destruction had already

Figure 12.1 Harborside road being dismantled in Zigui

begun months earlier. One apartment building, like a giant dollhouse, was missing the outside walls of several floors but there were lights on a floor above, and you could see a solitary figure moving about. The next block was completely dark, except for a family sitting in their house watching TV with the doors open, their dogs wandering in and out, just as they would in any town on a hot summer's night, but here surrounded by the vacant houses of their former neighbors, like the survivors of a plague.

As the new cities began to rise above the gorges and the old ones to fade, I took long walks through familiar places, thinking that each time might be my last before they were obliterated. In Wushan, the pink and green skyscrapers spiking up from the ridge of the tallest mountain, some 3,900 feet above the Yangtze, looked from a distance like a holiday resort, but the grandeur of the mountains, with their peaks barely visible through the mist and the specks and streaks representing people and footpaths far below, is somehow diminished by these new and angular shapes. This is one of the most jarring visual results of the new construction—the mountains are just as high, but, with the ridges lined with man-made shapes, the sense of the dominance of nature and its power is gone.

After the city was laid out and the major buildings completed, I took a cab to the top of Wushan to see the new town close up. Looking up from the semi-finished streets, it was not unattractive. The city government headquarters and the Communist Party compound tower over the city and river. On the main street are a teachers' training school, a senior middle school, apartment blocks, and a few odd bits of architecture, like the tiny Greek revival building said to be a future branch of the Bank of China. Construction workers live in temporary tin huts on the periphery of the new town hall. Near the shacks, children run back and forth and adults play cards by the edge of the road. According to a local taxicab driver, the apartments cost new residents about 90 yuan per square meter, depending on the location and floor level, a relatively low price because the expense is shared by the buyer and his employer or one of the government agencies. The driver was pleased with his new accommodations, although annoyed that the city would provide only 200 yuan for moving expenses (about U.S. $30), not nearly enough to transport an entire household up the mountainside, even with his cab.

A few hours after my visit to the new city, I was met on the gangplank to my ship by a delegation from the local Public Security Bureau, the agency, along with the police, responsible for maintaining law and order in China. The group included their specialist in dealing with foreigners and the problems they create, and my talkative cabdriver. They had come for me and my film, and hopefully a chat with the ship's political commissar about my behavior. According to them, the new town was not open to foreigners, even if officials in Beijing said it was, nor was it attractive, nor could it be photographed by foreigners. It would be beautiful in three years and I could come back and look at it then. Meanwhile, they would like my film, a demand made all the more annoying because the cab-driver had memorized every picture I had taken and had recited the locations with zeal. I thought of retreating to my cabin, locking the door, and waiting to see if we set sail as planned, but to avoid the potential awkwardness of involving the ship's officers, I decided to give in. The political commissar was fortunately busy with his best employee of the year awards and could not be disturbed, so I climbed into the public security car with my escorts and drove to a nearby camera shop so they could develop my film.

There the owner set to work washing my film so they could destroy any pictures of the new town and return my remaining photos. The public security officers and I sat on the shop's wooden benches and made polite conversation, Chinese style. I learned that one official and I were both born in the year of the horse and he was only three weeks older than I. The other man and I had met before, during a previous trip, and under the circumstances counted as old friends. The images eventually emerged from the film and everyone in our little group looked them over and complimented me on my photos before they cut them to shreds. The cabdriver was more excited about this than anybody and gave a running narrative. "We came to the new school and I said it's a school, and she said it's big and is it a high school and I said no, it's a normal school, and then she said is this an apartment building and I said it was, and she asked do you know where you'll be living yet and I said I did. . . ." A few negatives of high-rise buildings that could have been anywhere and some views of the river below escaped their scissors. The public security team then returned me to my boat, telling me they welcomed my return some other time.

I asked other people in town and in Chongqing why they thought the officials were so agitated, for such a superficial visit allowed no time to see anything more than a new city which looked much better than the old. Everyone gave the same answer—that it was another example of local fear and control, the heightened and ever-present anxiety that something bad would be revealed to the outside for which county cadres were or would be held responsible, and the small town need to show that local authorities make the rules about who can do or see what, regardless of regulations made in Chongqing or Beijing. A year later, back in Wushan, I was on the wharf, saying good-bye to someone on a departing ship when the dockhands struck up a conversation. They asked me if I knew about the foreigner who had come here last year, the one who had gone to Columbia University. She had gone up the hill and taken pictures, and the Public Security Bureau took them away to protect the safety of the county seat. They wondered if she would be back. I said I didn't know.

A large section of the road above the Wushan harbor and the buildings near it collapsed during the floods of 1998. These have never been rebuilt and much of the waterfront is a nearly impassable expanse of mud striated by deep crevices, now used by the armed police to practice wrestling and combat maneuvers. On the steep incline above what remains of the road is an unmoving waterfall of garbage, apparently there since the waters receded. One of the casualties of the landslide was the building that housed the Catholic church and the surrounding neighborhood, where the street now ends in a concrete cliff. I was confused the first time I went back, suddenly disoriented by the absence of a road I was sure had been there. As I tried to figure out what was wrong, a family in the house next to the space where the church had been invited me in out of the heat. I sat for a while with the six members of the three-generation family that lived in the old three-room house, and watched TV and drank tea next to the electric fan. They asked me questions about the United States and told me that no one had been hurt in the landslide, but that as a result of it, everyone who had lived nearby had moved away ahead of schedule. When I left, the housewife asked me to come back and visit the next time I was in town, and then she caught herself and said, "But we'll all be gone. You won't be able to find us."

Below, near the remains of the buildings that slipped down the hill, a woman selling water, film, and umbrellas told me that the town had decided there was no point in cleaning up anything but the dangerous rubble after the landslide since it would all have to be removed later anyway. Though her farmhouse on the edge of the city would be lost, she was looking forward to moving. With three children, a son who was a long-distance truck driver, and two daughters, one a peasant, the other a traditional doctor, her family was doing well, an example of the economic mobility that is possible for successful peasant families. All three children had been assigned housing in the new town, and the daughter who farms was already in a new apartment allocated to her husband by his work-unit. The shopkeeper, who is in her seventies, will move in with her son eventually, but for the time being she prefers to go back and forth between her temporary quarters on the dock and the house where she has lived much of her life. She likes sitting and talking to people as they get on and off the boats and watching the world come and go, and her children have turned out well. The future looks good.

Another dockside vendor told me that her son was going take the senior high school examinations in a few days. She and her husband were very nervous about this, but if the boy did not pass, they would send him south to one of the factories in Shenzhen or some other new industrial city to work in a factory where, they hoped, he could get some technical training on the job. The woman's own education had ended when she was nine, after her mother died and she had to leave school to keep house and take care of the family. With four years of schooling, she can more or less read, but not write. She said her son was a good student, and the most important thing was to get him off the land. She wanted him to have a good education, but the factory route would be all right, so long as he never had to be a peasant again.

Everywhere in the gorges there is a sense of urgency about education, of more opportunities for those people with a college degree or technical training and fewer for everyone else. In the countryside, many children, particularly girls, spend no more than a few years in primary school. In the reservoir area, about 65 percent of students go on to junior middle school, and most students in county towns and adjacent areas who live within a few miles walk of a school and have parents who encourage their education will complete the ninth grade. Graduation from junior middle school is a key dividing point, the minimal educational level necessary to get factory work in a big city, and enough to open up possibilities beyond construction or selling soft drinks or cigarettes on the street. Senior middle school (grades ten to twelve, the equivalent of American senior high school) separates students who will go to the fields and factories, or perhaps into new service industries, from those who may have a chance at a higher education and better long-range prospects. About half of the ninth grade graduates will go on to senior high school in Fengjie, slightly fewer in Wushan, more in Wanxian, and many fewer in poorer interior counties. In the countryside and small towns, the odds of getting into college or a technical school frequently determine the estimated value of education after the ninth grade, for unless a student is accepted by a school with a good record of success in the national college entrance examinations, parents often see no point in further schooling for young men and women who could instead be contributing to the family income.

The mass relocation of the population, which includes over 200,000 students from primary school to college, presents enormous logistical and educational challenges. The expansion and upgrading of schools in the reservoir areas is a key focus of the resettlement plans, but as with everything else, funding is inadequate compared with needs. Some assistance has come from the outside. A new Sino-Canadian primary school recently opened in a migrant district of Chongqing, and Tsinghua University in Beijing, China's equivalent of MIT, has helped sponsor and fund a technical high school, but the main financial and educational burdens lie with the local communities. In addition to the ongoing struggle to improve basic education in this area, there is now the problem of how and when to relocate approximately 600 existing schools in communities in which the residents, including children, are scheduled to move at different times depending on their location. This is complicated when one school may serve a large area whose inhabitants may be moving over a two- or three-year period. An even more difficult problem is the education of children moving farther away. The issues vary depending on where they come from and where they are going. In the town centers, it is mostly a question of orchestrating the arrival of classmates and teachers in new buildings at the same time, but for children from the countryside, it is much harder. Village students moving to new townships must adjust to being thrown together with thousands of strangers in large apartment blocks, and others relocating to interior rural areas may face educational opportunities even more limited than those they leave behind. For families settling in distant parts of China where Mandarin may be understood both by the exiles from Sichuan and their new neighbors, but is not the language of daily use for either, the move presents other problems and is almost always more difficult.

Wanxian's Longbao district was until recently a vast hilly stretch of fields and farmhouses. It is now tightly packed with white ten-story apartment complexes. I visited the Hongguan Primary School, established in 1997 for migrant children in this district with a grant from the immigrant bureau. The first school building, located elsewhere, collapsed in an avalanche and the 800 students had to be taught in makeshift classrooms in people's houses. The reconstructed school has since become a model, though one that may not be attainable, for other schools in the region. Of the 1,700 students, 300 are boarders, most of whom come from rural parts of the district too far away for a daily commute. Others are city children who are orphans or whose parents have jobs that make it impossible to be home at regular hours. The dormitory is comfortable and the classrooms spacious. There are computers, English lessons starting in the second grade, and a strong music and arts program.

The principal, Mrs. Li, has been a teacher or principal for over thirty years. She told me that the school has become a magnet for migrant families, with the result that households assigned to other parts of Wanxian try to move here or live with relatives in order to register their children in this school. Like the old Children's Palaces, the after-school programs often viewed by tourist groups, Hongguan is a showpiece for foreign visitors, but it also a place where real educational opportunities exist for children who have just been uprooted. I asked Mrs. Li whether her pupils had much trouble adapting to the large school and city life. She said that, in general, young children had few problems, though it was not always easy for their mothers and fathers. For children entering the school at a later age, in the fourth or fifth grade, it was harder. Coming from small village schools, they lagged behind the other pupils. Still, with special attention, most students catch up in the first year, mainly because they and their parents are so motivated. Almost all the peasants who come to the city, she pointed out, do so in part because they want their children to receive a kind of education unobtainable in the countryside.

Although the government has made education a priority in the reservoir area, funding remains a major problem, both for showcase schools like Mrs. Li's, and other institutions. Not only have state-owned factories been told to make themselves self-supporting and profitable, but so have the schools and every other kind of community organization. In addition to running the academic program, Mrs. Li must also oversee the investment of funds contributed by the government and is responsible for finding ways for the school to make money. The school has had to set up small businesses, and on the weekends women work in empty classrooms doing piecework assembly of small lantern-like items; plans are underway for a small machinery factory on the grounds. Even with outside advice and assistance, the burden of selecting and managing new economic ventures is a heavy one and, for smaller schools with little initial funding, can be almost impossible. Funding is grim in both Wushan and Fengjie and cannot keep pace with the number of new schools that must be built and staffed to educate children moving into the new towns and reclaimed rural areas. According to newspaper reports, Fengjie county borrowed 67 million yuan for education between 1996 and 2000 and is now finding it difficult to make payments on the loan. In Wushan, the situation is even more difficult, with an estimated lack of almost three-quarters of the funds needed to complete construction of hundreds of schools.

Other cultural and religious institutions are also beset by financial difficulties. Buddhist temples and Christian churches are among those struggling to find ways to finance their new buildings and programs. Although their moving and rebuilding expenses will be partially paid for with compensation funds, and in some cases with additional resources designated for cultural sites, these are not enough to cover the total cost of relocation and new operating and program expenses. Churches, the Catholic church in particular, have launched well-organized international appeals for assistance. A detailed description of the needs of each of eight churches in the six towns belonging to the Wanxian diocese is now available on the Internet. As a means of supplementing their incomes, most churches are planning to set up hotels and restaurants for out-of-town parishioners and

Figure 12.2 Construction workers in Zhongxian

tourists. In Fengjie, the women in the two-hundred-year-old Buddhist temple on the road overlooking the harbor said the temple in the new town will be even better, while the caretaker of the Protestant church down the street seemed less clear about its future.

In each town, finances drive the future. In the past, during the years when most town residents were employed by state-owned enterprises, Fengjie's relatively educated workforce and its reserves of coal, salt, and sulfur gave it advantages that other river towns lacked and made it reasonably prosperous by local standards. With the government's decision in the 1990s that state-run businesses must become profitable or close down, the area has come on hard times. Of the eighty-one such enterprises, which kept at least one or two members of almost every family in this town of 30,000 employed, as well as providing subsidized housing, medical care, and other benefits, only two were profitable by the year 2000. The other seventy-nine have undergone major reorganizations or have been closed. Outdated cement and fertilizer factories have been updated with the goal of finding new markets for a better product, with limited results, and attempts to attract outside investment, like the Tropicana orange juice factory in Zhongxian, have met with little interest. Some large new businesses, like the three-star hotel in downtown Fengjie, have opened, along with scores of tiny privately owned shops and services. The hope has been that by getting the larger businesses up and running in the old cities before the relocation, it would be possible to move already functioning enterprises and employed people. Thus, as a representative of the government travel service explained, the new hotel was not an illegal new structure below the high-water mark but a start-up enterprise designated for the new migrant zone.

Despite such efforts to restructure and revitalize the economy, officials are uncertain about Fengjie's future prospects. As elsewhere in China, successful reorganizations have made factories more efficient, but they have also led to widespread layoffs. As the factories have ceased to be reliable sources of employment for young people, most have sought work in the new shops and construction or have left on two-year factory contracts elsewhere. For young men and women with average educations, there are fewer jobs than before, and for those who lack the drive and skills to make it on their own some other way, the future can look bleak. Farming is hard here, and with land taxes and smaller families, it is difficult to make ends meet without some family members employed off the land. For the time being, thousands of men are needed to work on road building and other construction projects, to shovel coal and break rocks, and to haul and carry. The question is whether new industries will open or can survive once there are better roads and communications, or if, as many fear, a better infrastructure will merely facilitate a faster trip from Chongqing to Yichang, making it easier to bypass the towns on the way. As one former official in Fengjie put it, "It is not just a question for Fengjie. What happens to the economy and the people as the state-owned enterprises close is one for all of China."

Early one morning, I made the long walk from the center of Fengjie through the outskirts of the city and into the countryside to the temple at Baidicheng, knowing I might not have the time or opportunity to do so again before these neighborhoods were underwater. The shops and stalls in the center of town were filled with shimmering plastic shoes, from baby shoes that squeak and light up to rubber camouflage sneakers and indestructible rhinestone-studded see-through plastic high heels, far more shoes than the population could ever buy. This abundance is one of the signs of an economy based far more on hope than reality, where despite their short lives, shops open every spring and are gone by midsummer, their proprietors looking anew for success with some other low-priced and easy-to-obtain necessity. Unlike most of the other river towns, Fengjie is not just a place that people leave, but one they come to as well, a place where migrants from downriver, as far away as Jiangsu and Jiangxi provinces, find small commercial opportunities less competitive than in more prosperous cities. Bakeries are plentiful, filled with bread and elaborately decorated cakes that are allegedly a legacy of the years when Western merchants and missionaries made frequent stops here.

I went past the city square where the morning market was overflowing with purple cabbages, beans, and peppers. Badminton players, martial arts students with their masters, tai chi groups, and marching troupes of "women over fifty" in red and white uniforms with drums and fans in constant motion compete for space yet avoid collisions. Except for the badminton players, who are all ages, ranging from children to their grandparents, all activities are segregated by age and sex. Elderly men play Chinese croquet, hitting ceramic balls across two large sections of the square marked off by string. They stand deep in concentration as they swing their mallets, a crowd of onlookers watching carefully. Other old men sit on benches below the public notice boards and swing their birdcages back and forth to air them or hang them on low-hanging electric light wires.

Up the hill from the square, you pass small shops and hairdressers, a store where business is not so good that enlarges snapshots on a computer, and then go through a back alley between old stone houses. Dumplings and fried pancakes are for sale in doorways, and children stop on the steps for a quick breakfast before making their way to the middle school just ahead, the

Figure 12.3 Motorcycle taxis near the Yidou Gate in Fengjie

best one in Fengjie, the one most parents hope their children will attend. The road then leads into the rural outskirts of the city and the banks of the Meixi River, which the local people call the Shu Guo River, after the kingdom of Shu, located here in the third century AD during the Three Kingdoms period. The banks are terraced into small fields from the road to the riverbank, and on the water men and boys in small boats sit quietly and fish. Peasants, coal shovelers, day laborers, and office workers come into town on foot along this road. Others make the trip by motorcycle or bus, often bringing their children to the city schools. On one side of the road are houses and a few caves, on the other, fortunetellers set up shop in the early morning shade, hanging their banners for numerology, palm reading, or phrenology on trees and electric poles. The bridge over the Shu Guo River was completed in 1994, making it possible to walk or drive to Baidicheng from Fengjie rather than go by boat, as had been necessary previously. A second bridge over the Caotang River, connecting Baidicheng to the Laoguanmiao section of the north bank of the mouth of Qutang Gorge, was built the following year. Both will be below the new high-water mark and will have to be taken down. As Lin Junwei, an official in Fengjie said in the early 1990s, when writers from Beijing criticized the construction of new buildings below the projected new water level, "You reporters have criticized us for building under the level to be submerged, but is it really so simple? Even excluding natural disasters along the way, it will still take twenty years to build the Three Gorges Dam. During this time, a generation has to survive. We have more than a million square meters of disintegrating and dangerous housing here. We have to take care of the people who live in these houses in the wind and rain. Furthermore, we have a population growth rate of 1.2 percent. Where do you want these children to live? In the open air?"[4] New building has continued in the old towns even as the new cities go up.

After making the loop north of the city to cross the Shu Guo River, you head south on the opposite bank toward the Yangtze again. The new bridge over the river forms a high arc behind the old one, a common sight now along the Yangtze, where few bridges were built on the tributaries until the early 1990s, most below what will be the new water level. A sign that is in every town, *yimin di erqi wei*, roughly translated as "the second migration level," stretches across the bank of the river. This refers to the second rise in the water level, scheduled for 2003, before which the second group of migrants must move. (The first group of several thousand migrants moved in 1997, at the time of the diversion of the Yangtze; the third will be relocated by 2006.)

Figure 12.4 Fields on the bank of the Meixi River in Fengjie, to be submerged in 2003

Some farmhouses along the road are nestled close against one another in groupings of eight or ten, others stand alone in the midst of large and luxurious fields where husbands and wives water their crops by hand, first drawing water from irrigation ponds in buckets hanging from the poles on their shoulders, and then ladling it out on to each plant. In the early hours of the morning many people are already at work, but some make a slower start, lingering on the flat roofs that are the center of daily life here. Whole families perch together, eating breakfast, washing the dishes, washing their hair, scrubbing clothes, brushing their teeth. Clothes are mended and hung to dry, grain threshed, and vegetables chopped. Women shout back and forth from roof to roof and watch their children in the roads and fields below. The banks of the Yangtze sound like a battlefield, with constant distant and nearby explosions. The uniform high-rise buildings, described even by their architects as *yiban*, or ordinary, are visible in the distance everywhere, and a new highway is being carved out of the mountainside. The Pinghu Pagoda, a Fengjie landmark since the tenth century and the highest structure here until recently, now looks small and overwhelmed. It, too, will soon disappear.

As I went along, people volunteered miscellaneous pieces of information. One man pointed to a spot where everyone had gathered to watch one of the biggest Japanese bombing raids on the city. I asked if anyone was hurt. He said not so many. "Almost no one had seen airplanes before the war so we thought it was fun to go up to the top of the hills to see them swoop down and to scream at them." Although countless foreigners visit Baidicheng, and Fengjie has had at least one resident foreign-language teacher, outside travelers do not often come into the city. As I walked along, I learned, as has happened before, that I was the only foreigner in town that week. People stopped to ask me where I was going, why I was alone, and why I was walking, since no one does that anymore. They told me about the German explosives experts who had been there for several weeks some months before and made the workers stay late to finish the work, even though it was time to go home. A woman overseeing a group of men taking down her building by hand, brick by brick, explained that her family was due to move to Fujian in three months and they couldn't take the house with them, so they were selling the bricks now. She told me her teenage son was at home and had nothing to do these days. He should have been taking the senior middle school entrance examinations, but since they were going to Fujian, there seemed no point. He would probably have to go to work anyway, or maybe he would go to a school for migrants or learn a trade. Several hundred people from this part of Fengjie would leave together, and she would find out in a few months whether they would continue to farm or if there would be new opportunities in factories. Other Fengjie neighborhoods have been assigned to towns in Zhejiang province. Villagers from Wushan, who had been assigned to Xinjiang before it became clear that the difficulties there were too great, are heading to Guangdong province instead, and thousands of residents of Zhongxian will make new homes in Shandong province.

In the small towns along the Yangtze, life is a peculiar mix of old and new, a sudden hurtle from a past resistant to change to an unknown future. Dozens of coal transfer and processing stations line the banks, where coal is brought from mines in the interior and dumped down the hillside to the edge of the river, and then is shoveled, bucket by bucket, into coal barges headed for Chongqing and Wuhan. Just as young men in Wushan do, laborers in Fengjie start at six in the morning, walking in from the countryside along with the farmers who sell their produce in the square. Some of these young men come from parts of Fengjie that will be relocated because of the dam, but most are from interior villages, a one- or two-hour walk away, that will not be affected by the change in the water level. They come to Fengjie between planting and harvesting because there is no other work in their home villages. Throughout the morning they shovel and rest and then shovel some more. Once in a while you see a young man hunched atop a coal heap above the river, cell phone in hand, and wonder whom he is calling. The workers have little education and are from poor farming families in the countryside, but cell phones have become a status symbol among the young, even here. No one is sure what will happen to them and their families as the economy changes. Planners envision new cities whose streets are not lined with young men in brilliant-colored undershirts shoveling coal, cities with less need for coal altogether, and deeper tributaries that will make it possible for barges to sail directly to the mines to pick up the coal at its source, eliminating the repeated truck and bucket transfers. But no one has a clear vision of what the people who have filled these jobs will do.

Across from Baidicheng, the White King city, is Laoguanmiao, the site of the old signal station at the mouth of Qutang Gorge, where extensive archeological excavations begun in 1990 resulted in the discovery of the skeletons of a mammoth and a Neolithic man. The mammoth has been put together again. Along with a detailed history of the excavation and the region, it forms the main exhibit of an attractive museum opened in 1995. Both Baidicheng and Laoguanmiao will remain above the new water level, becoming two of a number of small islands along the edges of the new Yangtze, little humps of history, formerly part of the daily life of the people who used to live here.

In the museum garden, a life-size mammoth is about to charge from a flower bed. Employees sit in the shade looking down at another 175-meter sign and the newly added, and much larger, China Mobile cell phone advertisement on the cliff face of Kuimen Gate at the entrance of Qutang Gorge. Inside, beyond the main display, by a sign leading to further exhibits in the cellar, an elderly woman grabbed my arm. She said, "There's something frightening down there. We'll go down and look at it together," and we did, descending the stairs into darkness. Just as I assumed that this was simply another unattended room and wondered where the light switch was, there was an angry roar, and an illuminated dinosaur leaned forward and growled. A cave man stood next to a panda and reptiles, and primitive birds shrieked. The mammoth wagged his tail and ears. The old lady yelped and held tightly to my arm, and said we were done and would now climb back to the top of Baidicheng.

When we emerged into the sunlight from the prehistoric lair, she stared at me and said, "When I looked at you before, I said to myself, that must be a foreigner. Are you a foreigner?" I admitted to this, and she nodded in satisfaction and told me that of all the heroes of the Three Kingdoms era, people from the Three Gorges most revere Zhuge Liang, steadfast friend and advisor of Liu Bei, emperor of the Shu Kingdom. Before his death, Liu Bei entrusted

Figure 12.5 Garden of the Mammoth Museum in Fengjie

the care of his youngest son to Zhuge Liang, who then raised him as a righteous man. Outsiders, visitors from other parts of China and Japan, admire Liu Bei, Zhang Fei, and other characters from this period for their military and strategic skills. These matter less, my companion said, than Zhuge Liang's loyalty and devotion, and she thought it was a peculiar difference in perception between people here and other places. As we climbed the mountain, visitors asked her if we were together. She beamed and said with pride of ownership, no, I met her by the roaring tiger—he was terrifying. After several moments of confusion, she would clarify the location—in the museum, in the elephant museum, on the hill—and then continue to muse about Liu Bei and his decision to entrust his youngest son to his friend, and how well he had brought him up.

As much as anything else, it is this day-to-day connection between past and present that is part of the living culture in the gorges that will be lost when they are flooded. There is not enough time or money to save all the artifacts and sites of importance along the Yangtze, and even fewer resources are available to identify and record the surviving traditions of the region—the indigenous music and arts, religion, and local customs related to marriages and funerals. In the mid-1990s, a team from the Central University for Nationalities in Beijing, working with anthropologist Zhuang Kongshao, made what was to be the first of a six-part film series covering a wide range of topics related to local traditions and beliefs, but further funding was not available and the project had to be abandoned. The stories of the Three Kingdoms are familiar to everyone in China and thus the related relics and temples are of concern to the general public. Most people remember the carvings at White Crane Ridge (the majority of which will be submerged nonetheless) and famous temples, such as the one housing Qu Yuan's spirit and clothes that will follow the residents of Zigui to higher ground. Few know or care that the Tujia, who live near the Hubei-Sichuan border, still sing, dance, and spin in circles around enormous red coffins at their funerals, rites that have their origins thousands of years ago in the ancient indigenous Ba culture. Even fewer are aware that traditional songs in the valleys of the gorges entreat the ground to grow while mountain dwellers sing mostly about gods and nature, or that at least since the Ming dynasty there have been families that have purposely isolated themselves in the peaks of the gorges, whose customs are entirely different from the farmers or fishermen below, and who live the way they do because they are completely free to do as they like, as long as they can eat.

When I have asked natives of the gorges if there are any particular characteristics that distinguish them from people in other parts of China, or differentiate, for example, the residents of Fengjie from those of Yunyang, or explain the centuries of conflict between Wushan and Zigui, they are generally puzzled by the question. Shanghaiese are known as clever, Cantonese for their business flair, Sichuanese and Hunanese for their fiery tempers, but no one can think at first of a common identifying feature or stereotype for anyone in the Three Gorges. One official told me that there probably was something unique to people in different parts of the gorges, but as he hadn't spent much time in them, he wouldn't know what it was. Another said that the girls in Fengjie are pretty. After thinking it over for a bit, almost everyone finally says that they share the ability to *chi ku*, to endure hardship.

Both officials and the general population mostly expect that this common trait of putting up with difficulty will be enough not only to get the inhabitants of the gorges through the move, but to sustain peasants relocating to harsher terrain through what may be years of difficulty. The fact that almost everyone in the reservoir area is ethnically Han and shares the same general traditions comes up frequently. The assumption is often made that when people from different places must suddenly live together, conflict or discomfort will result only if they are from distinct cultural groups, such as Moslems or hill tribes. As such a mix does not exist here, the expectation is that the broad cultural homogeneity of the gorges will quickly dispel any difficulties. Few officials I spoke with seemed to think that villagers accustomed to a social structure based on entire communities of people living their lives within sight of one another might find the transition harsh when these familiar foundations disappear. Chinese sociologists and historians, on the other hand, predict that for the older generation in particular, change may be hard. Elderly peasants who believe that the fundamental relationship between man and earth means that man belongs on the ground can feel permanently ill at ease when assigned to eight-story apartment buildings. Farmers accustomed to the well-thought-out placement of their houses now worry about the difficulties that may result from the endless rows of identical buildings. Unlike Hong Kong and Singapore, where careful attention is paid to the *feng shui* of new buildings to minimize negative influences, placement has not played a role in city planning here. The simple urbanizing change from single-family houses to vertical high-rises makes many residents anxious, and they wonder if they will be able to cook on their balconies when it is hot, and how they will talk to their neighbors if they are on top of each other.

Others point out that hundreds of thousands of Chinese peasants have made the transition from the countryside to the city and none of them regret it. Many of the travelers on the passenger boats and buses leaving Chongqing for towns along the river and interior counties like Kaixian and Liangping are on their way to visit their *lao jia*, their family's town of origin, a place they may have left for university or which their grandparents fled in wartime. On the new express bus from Chongqing to Wanxian, a clean, fast, and terrifying ride through the mountains and interior countryside, I have sat next to stockbrokers from Chengdu, officials from Chongqing, shopgirls, and factory workers. All claim a deep attachment to the region, and none would ever live there again. They send money to help parents and grandparents build new houses or invest in businesses, but their goal for brothers and sisters and nephews and nieces is to get them out.

Travel companies are now designing new itineraries for a new river. They say that there will be trips up the widened tributaries of the Yangtze, but as the reservoir itself will be less interesting, travel will focus more on the mechanics of the dam, much like Panama Canal tours. The only part of the gorges that will not be substantially changed by the Three Gorges Dam is the short section of the river between the entrance to Xiling Gorge, just past the Gezhouba Dam and Sandouping. Two famous historical sites are located within this stretch, *San You Dong*, or Three Travelers Cave, at the mouth of the gorge near Yichang, and Huangling Temple, located just east of the new dam site. One of the oldest structures in the Three Gorges region, Huangling Temple is dedicated to the ancient Emperor Yu for creating the gorges and saving China from the floodwaters. First built in the Spring and Autumn Period (770–476 BC), it was enlarged in the third century AD by Zhuge Liang (who raised Liu Bei's son) and contains hundreds of inscriptions related to water management and control. Three Travelers Cave takes its name from the visit of the poet Bai Juyi, his younger brother, and a third poet friend, who in 819 wrote:

> We moored our boat under the cliff and cleared the tall wall so that we could go forward. Where the rocks were too slippery, we used ropes in climbing and did not reach the summit until we had tried four times. We looked around from the top and found no human habitation. Waves beat against the stones, which, though submerged in the water, could be seen clearly. The froths were in the shape of pearls and jades, offering a marvelous sight. We stayed at the spot enjoying the landscape from the early afternoon until late in the evening, and still we were reluctant to go. Since the three of us were the first to visit the cave, it was named the Three Visitors Cave.[5]

One late night in my younger years, when the *Kun Lun* docked by the cliff, very close to where Bai Juyi and his brother stopped almost twelve hundred years earlier, a friend and I quietly disembarked. A park with steps now surrounds the entrance to the cave, so we did not need ropes to reach its opening, but we still had to scale a high ornamental iron fence to get to the entrance in the middle of the night. The multicolored disco-like lights that illuminated the cave during working hours were off, and only the dim glow of emergency bulbs made it possible for us to find our way though the first chamber and into the large cathedral-like opening. More than forty stone inscriptions, including Bai Juyi's carvings from the eighth century AD and the comments of the Tongzhi emperor (1856–1875), written a thousand years later, are still discernable on the cave walls. The Tongzhi emperor's inscription states, "Time is not as important as place, place is not as important as harmony with the people, and harmony with the people is what guarantees the stability of place to everyone."[6] The quote expresses the ancient Chinese ideology that the stability of nature, the well-being of the people, and beneficent rule are all intertwined, a belief still commonplace along the Yangtze. The other side of this conviction is that it is important to keep nature under control, as the Three Gorges Dam, the largest of all such projects, is supposed to do. As land and nature are transformed in the gorges, people from Chongqing to Yichang, and in new far away migrant communities, now wait, in resignation, hope, or anger, for whatever comes after the deluge.

NOTES

INTRODUCTION

1. Bai Juyi, "On Being Removed from Hsun-yang and Sent to Chung-chou," *One Hundred & Seventy Chinese Poems*, translated by Arthur Waley (New York: Alfred A. Knopf, 1923), p. 219.

CHAPTER 1

1. Isabella Bird, *The Yangtze Valley and Beyond: An Account of Journeys in China, Chiefly in the Province of Sze Chuan and Among the Man-tze of the Somo Territory* (London: John Murray Press, 1899; reprint, London: Virago Press, 1985), p.153.
2. Ibid., p. 273.

CHAPTER 2

1. Gong Li, *Daning River* (Wushan: unpublished manuscript).
2. Ban Gu, *Han shu* (The Book of the Han Dynasty) (Beijing: Zhonghua shuju, 1975), chap. 28, *Dili zhi* (Geography), pp. 1597, 1603.
3. Lyman P. Van Slyke, *Yangtze: Nature, History, and the River* (Reading, MA.: Addison-Wesley, 1988), p. 134.
4. Du Fu, "Bazhentu" (Eight Fortresses). Translated by Deirdre Chetham and Shuxi Yin.

CHAPTER 3

1. Lu Yu, *The Old Man Who Does As He Pleases: Selections from the Poetry and Prose of Lu Yu*, translated by Burton Watson (New York: Columbia University Press, 1973), pp. 117–18.
2. Sergei M. Tretiakov, *A Chinese Testament: The Autobiography of Tan Shih-hua, as Told to S. Tretiakov* (Reprint, Westport, CT: Hyperion Press, 1978), p. 7.
3. Yangtze Valley Planning Office, ed., *The Three Gorges of the Yangtze River: A Guide for Tourists* (*Sanxia daguan*) (Beijing: Shuili dianli chubanshe, 1986), p. 212.
4. Ibid., p. 218.
5. Fan Chengda, "*Lao yu geng*" (Yu Cultivation), *Shihu jushi shiji* (Poems of Fan Chengda) (Shanghai: Shangwu yinshu guan, 1929), vol. 16, pp. 6, 7.

6. Lu Yu, *The Old Man Who Does As He Pleases*, p. 116.
7. Peng Zunsi, *Shu bi* (Sadness of Shu) (Changsha: Shangwu yinshu guan, 1939), p. 54. Translated by Xiaoping Chen.
8. Shen, Congwen, *Recollections of West Hunan*, translated by Gladys Yang (Beijing: Panda Books, 1982), pp. 168–69.
9. G. R. G. Worcester, *Sail and Sweep in China* (London: H.M.S.O., 1966), p. 127.
10. Lyman P. Van Slyke, *Yangtze: Nature, History, and the River* (Reading, Mass.: Addison-Wesley, 1988), p.118.
11. Cornell Plant, *Glimpses of the Yangtze Gorges* (Shanghai: Kelly & Walsh, 1921), p. 13.
12. Yangtze Valley Planning Office, ed., *The Three Gorges of the Yangtze River*, p. 286.
13. Ibid., p. 295.
14. Faxian, *A Record of Buddhistic Kingdoms Being an Account by the Chinese Monk of His Travels in India and Ceylon (A.D. 399–414) in Search of the Buddhist Books of Discipline*, trans. by James Legge (Oxford: Clarendon Press, 1886), p. 112.
15. Worcester, *Sail and Sweep in China*, p. 10.
16. G. R. G. Worcester, The *Junks and Sampans of the Yangtze: A Study in Chinese Nautical Research* (Shanghai: Statistical Department of the Inspectorate General of Customs, 1948), p. 464.
17. Ibid., p. 463.
18. Plant, *Glimpses of the Yangtze Gorges*, p. 16.
19. Lu Yu, *The Old Man Who Does As He Pleases*, pp. 105–6.
20. John Hersey, *A Single Pebble* (New York: Vintage Books, 1989), pp. 58, 60.
21. Mencius, *Mencius*, translated by D.C. Lau (Harmondsworth: Penguin, 1970), p. 51.

CHAPTER 4

1. Li Bai, "The Sichuan Road," *Poetry and Prose of the Tang and Song*, trans. by Yang Xianyi and Gladys Yang (Beijing: Panda Books, 1984), pp. 19–22.
2. Thomas W. Blakiston, *Five Months on the Yang-Tsze, and Notices of the Present Rebellions in China* (London: John Murray Press, 1862), pp. 147–49.
3. Alexander Hosie, *Report by Consul-General Hosie on the Province of Ssu-chuan* (London: Printed for H.M. Stationery Office by Harrison and Sons, 1904), p. 28.
4. Robert J. Davidson and Isaac Mason, *Life in West China: Described by Two Residents in the Province of Sz-Chwan* (London: Headley Bros., 1905), p. 148.
5. Ibid., p. 36.
6. William Gill, *The River of Golden Sand: The Narrative of a Journey through China and Eastern Tibet to Burmah*, vol. 1 (London: John Murray, 1880), p. 249.
7. Isabella Bird, *The Yangtze Valley and Beyond: An Account of Journeys in China, Chiefly in the Province of Sze Chuan and Among the Man-Tze of the Somo Territory* (London: John Murray, 1899; reprint, London: Virago Press, 1985), p. 151.
8. Ibid., pp. 154, 155.
9. John Hersey, *A Single Pebble* (New York: Vintage Books, 1989), p. 37.
10. Bird, *The Yangtze Valley and Beyond*, p. 147.
11. Wang Jian, "Shuifu yao" (A Boatman's Song), in *Tangshi xinxuan sanbai shou* (Newly Selected 300 Poems of the Tang) (Beijing: Renmin wenxue chubanshe, 1980), p. 274. Translated by Shuxi Yin and Deirdre Chetham.

12. Archibald John Little, *Through the Yang-tse Gorges, or Trade and Travel in Western China* (London: Sampson Low, Marston, Searle, & Rivington, 1888), p. ix.
13. Sir Edward Hertslet, *Hertslet's China Treaties. Treaties, &c. between Great Britain and China; and between China and Foreign Powers; and Orders in Council, Rules, Regulations, Acts of Parliament, Decrees, &c. Affecting British Interests in China. In force on the 1st January, 1908* (London: Printed for His Majesty's Stationery Office, by Harrison and Sons, third edition; revised, under the superintendence of the librarian of the Foreign Office, by Godfrey E. P. Hertslet, with the assistance of Edward Parkes, 1908), vol. 1, p. 77.

CHAPTER 5

1. Sergei M. Tretiakov, *A Chinese Testament: The Autobiography of Tan Shih-hua as Told to S. Tretiakov* (Reprint, Westport, CT: Hyperion Press, 1978), p. 71.
2. Meyrick Hewlett, *Forty Years in China* (London: Macmillan, 1943), p. 94.
3. Robert A. Kapp, *Szechwan and the Chinese Republic* (New Haven, CT: Yale University Press, 1973), p. 10.
4. Tretiakov, *A Chinese Testament*, p. 165.
5. "Gen. Yang Favors Swimming for Women, Paved Streets and Universal Cleanliness," *The New York Times*, August 20, 1933, sec. 4 (Editorial Section), p. 3. Cited by Kapp, *Szechwan and the Chinese Republic*, p. 29.
6. Tretiakov, *A Chinese Testament*, p. 197.
7. Kemp Tolley, *Yangtze Patrol* (Annapolis, MD.: Naval Institute Press, 1971), pp. 82, 83.
8. Ibid., p. 102.
9. Ibid., p. 198.
10. Ibid., p. 118.
11. Kapp, *Szechwan and the Chinese Republic*, p. 41.
12. Ibid., p. 40.
13. Ibid., p. 37.
14. Graham Peck, *Through China's Wall* (Boston: Houghton Mifflin Company, 1940), p. 170.
15. Kapp, *Szechwan and the Chinese Republic*, p. 87.
16. Ibid., p. 88.
17. Ibid., p. 71.
18. Theodore H. White and Annalee Jacoby, *Thunder Out of China* (New York: William Sloane, 1946), pp. 9, 10.

CHAPTER 6

1. Theodore H. White and Annalee Jacoby, *Thunder Out of China* (New York: William Sloane, 1946), p. 59.
2. Ibid., p. 11.
3. Dick Wilson, *When Tigers Fight* (New York: Viking Press, 1982), p. 175.
4. G. R. G. Worcester, *Sail and Sweep in China* (London: H.M.S.O., 1966), p. 102.
5. White and Jacoby, *Thunder Out of China*, p. 274.
6. Sun Yat-sen, *San Min Chu I: The Three Principles of the People* (Chungking: Ministry of Information of the Republic of China, 1943), pp. 462, 463. Cited in Yin Liangwu, "The Long Quest of Greatness: China's Decision to Launch the

Three Gorges Project" (Ph.D. diss., Washington University in St. Louis, 1996), pp. 78, 79.

7. Li Bin, ed., *Zhongguo Sanxia jianshe nianjian* (China Three Gorges Project Construction Yearbook) (Beijing: Zhongguo Sanxia chubanshe, 1995), p. 261. Cited by Yin, "The Long Quest of Greatness," p. 93.
8. Zhu Dajing, "Shinian lai zhi dianli shiye" (Electricity Industry in the Past 10 Years), pp. 388–408. Cited by Yin, "The Long Quest of Greatness," p. 136.
9. Yin, "The Long Quest of Greatness," p. 118.
10. John King Fairbank, *The United States and China* (Cambridge: Harvard University Press, 1948), p. 338.
11. Yin, "The Long Quest of Greatness," pp. 118, 119.
12. "First Five-Year Program for China's Postwar Economic Development" in China Oral History Collection, Related Papers, Box 39, Rare Book and Manuscript Library, Columbia University, pp. 113–42, 213–21. Cited by Yin, "The Long Quest of Greatness," p. 134.
13. Information Service, Bureau of Reclamation, Department of the Interior, "Background Information to Accompany Picture Sequence 'A Halter for the Yangtze,' Released on February 28, 1946." Denver, CO: Bureau of Reclamation Library. Cited by Yin, "The Long Quest of Greatness," p. 195.
14. Copy of Letter from Donald M. Nelson to President, the White House, Washington, March 19, 1945," with "OK, F.D.R." on the letter, Folder: China, 1943–1945, Box 2, Official File 150. New York: Franklin D. Roosevelt Library. Cited by Yin, "The Long Quest of Greatness," p. 236.
15. "United States Department of the Interior, Bureau of Reclamation, News Release, Wednesday, November 28, 1945," in "Drawings, Data, Correspondence, and Library Material, Yangtze Gorge Project, China." Denver, CO: Federal Records Center, National Archives and Records Administration. Cited by Yin, "The Long Quest of Greatness," pp. 281, 282.
16. Mao Ronglin, "YVA yu jinhou jianguo" (YVA and the Future Reconstruction of the Country), *Zhongyang ribao* (Central Daily News) (Nanjing), December 2, 1946, p. 11; and "YVA yu Zhongguo minzhu" (YVA and China's Democracy), *Zhongyang ribao* (Central Daily News) (Nanjing), April 17, 1947, p. 5. Cited by Yin, "The Long Quest of Greatness," p. 285.
17. Lewis Clark, Minister-Counselor of Embassy in Nanjing, to the Secretary of State, March 15, 1948, Subject: "Transmission of Text of Speech by Premier Chang Chun on Importance of Yangtze Valley Region and Need for Foreign Assistance," "Address by Premier Chang Chun, March 13, 1948," RG 59, Department of State, Box 4844, Decimal File, 1945–1949, 811.503193/3–1548, National Archives. Cited by Yin, "The Long Quest of Greatness," p. 316.

CHAPTER 7

1. Jung Chang, *Wild Swans* (New York: Global Flair, 1991), p. 148.
2. *Wushanxian zhi* (Wushan County Gazetteer) (Chengdu: Sichuan renmin chubanshe, 1991), p. 22.
3. Ibid., p. 24.
4. Chang, *Wild Swans*, p. 483.
5. Yuan Ran, "Fengyu lishu wan" (Wind and Rain in Pear Tree Town), *Nüwa, laozhai, Fengqingjie* (Short Fiction from the Three Gorges Area) (Hong Kong:

Wenguang chubanshe, 1993), p. 86. Translated by Sahoko Shiga and Deirdre Chetham.

CHAPTER 8

1. Zhou Weichun, ed., *Meng yuan Yichang* (The Dream Came True in Yichang) (Wuhan: Hubei renmin chubanshe, 1998), p. 60.
2. Shi Wenyin, *Sanxia: Yige kua shiji de meng* (The Three Gorges—A Dream Across the Centuries) (Guangzhou: Huacheng chubanshe, 1992), p. 81.
3. Zhou Weichuan, ed., *Meng yuan Yichang*, p. 62.
4. "Chairman Mao Swims in the Yangtse," *China Pictorial* (October 1966), pp. 1, 2.
5. Yangtze Valley Planning Office, ed., *The Three Gorges of the Yangtze River—A Guide for Tourists* (*Sanxia daguan*) (Beijing: Shuili dianli chubanshe, 1986), p. 46.
6. Hu Mingsi and Luo Chengzheng, eds., *Zhongguo lishi da hongshui* (Great Floods in Chinese History) (Beijing: Zhongguo shudian, 1992), vol.2, p. 305.
7. William Ferdinand Tyler, *Pulling Strings in China* (London: Constable & Co., 1929), p. 161.
8. *Along the Changjiang River* (Wuhan: Changjiang Branch of China International Travel Service), p. 30.
9. Cao Yinwang, *Zhou Enlai yu zhishui* (Zhou Enlai and Water Conservancy) (Beijing: Zhongyang wenxian chubanshe, 1991), pp. 14–16.
10. Central Intelligence Agency, Office of Research and Reports, *The Program for Water Conservancy in Communist China, 1949–61* (Washington, DC: Central Intelligence Agency, 1963), pp. 33–37. Cited by Liangwu Yin, "The Long Quest of Greatness: China's Decision to Launch the Three Gorges Project" (Ph.D. diss., Washington University in St. Louis, 1996), pp. 320, 321.
11. G. R. G. Worcester, *Sail and Sweep in China* (London: H.M.S.O., 1966), p. 128.
12. Cao Yinwang, *Zhou Enlai yu zhishui*, pp. 26–35.
13. Rewi Alley, *Man Against Flood* (Beijing: New World Press, 1956), p. 69.
14. United States Senate, Report and Staff Studies to the Committees on Interior and Insular Affairs and Public Works, Relative Water and Power Resource Development in the U.S.S.R. and the U.S.A. (Pursuant to S. Res. 248, 85th Congress, 2nd Session, Washington DC: U.S. Government Printing Office, January 4, 1960), pp. 132,133. Cited by Yin, "The Long Quest of Greatness," p. 334.
15. Li Zhisui, *The Private Life of Chairman Mao: The Memoirs of Mao's Personal Physician*, translated by Tai Hung-chao, editorial assistance by Anne F. Thurston (London: Chatto & Windus, 1994), pp. 166, 167.
16. Li Rui, *Lun Sanxia gongcheng* (On the Three Gorges Project) (Hong Kong: Mingliu chubanshe, 1998), p. 266.
17. Ibid., pp. 269, 270.
18. Jin Xiaoming, *Fengyu Sanxia meng: Sanxia gongcheng jishinian dajishi* (The Three Gorges Dream: A Chronicle of Several Decades of the Three Gorges Project) (Chengdu: Sichuan renmin chubanshe, 1992), p. 92.
19. Li Rui, *Lun Sanxia gongcheng* (On the Three Gorges Project) (Changsha: Hunan kexue jishu chubanshe, 1985), pp. 94–99.
20. "Guanyu Sanxia shuili shuniu he Changjiang liuyu guihua yijian" (The Opinions on the Three Gorges Key Water Control Project and Yangtze Valley Planning), adopted at the CCP Central Committee Chengdu Meeting, March 25, 1958, in Mao Zedong and Zhou Enlai, "Mao Zedong, Zhou Enlai guanyu

Changjiang Sanxia gongcheng de ruogan tanhua he pishi" (Some Talks and Instructions by Mao Zedong and Zhou Enlai on the Three Gorges Project in the Yangtze River), March 1958-November 1972, *Dang de wenxian* (Party Documents), no.1 (1993), pp. 46, 47. Cited by Yin, "The Long Quest of Greatness," p. 358.

21. Zhou Weichun, ed., *Meng yuan Yichang*, p. 58.
22. "The Yangtze Gorges Submit to Man," *China Reconstructs*, vol. 23, no. 9 (September 1974), p. 28.
23. Cornell Plant, *Glimpses of the Yangtze Gorges* (Shanghai: Kelly & Walsh, 1921).
24. "The Yangtze Gorges Submit to Man," *China Reconstructs*, vol. 23, no. 9. (September 1974), p. 30.
25. "The Hanchiang River Story," *China Reconstructs*, vol. 23, no. 6 (June 1974), p. 15.
26. "Chairman Mao Swims in the Yangtse," *China Pictorial* (October 1966), p. 3.
27. Lu Yuegang, *Changjiang Sanxia: Bange shiji de lunzheng* (The Three Gorges: Debate for Half a Century) (Beijing: Zhongguo shehui kexue chubanshe, 1993), p. 80.
28. Cao Yinwang, *Zhou Enlai yu zhishui*, pp. 50–52.
29. Mao Zedong, "Dui 'zhonggong zhongyang guanyu xingjian Changjiang Gezhouba shuili shuniu gongcheng de pifu' de pishi" (Instructions on 'The Approval of the Chinese Communist Party Central Committee on Construction of Gezhouba Water Conservancy Key Engineering Project in the Yangtze'), p. 48. Cited by Yin, "The Long Quest of Greatness," pp. 386, 387.
30. Yin, "The Long Quest of Greatness," p. 388.
31. Lu Yuegang, *Changjiang Sanxia: Bange shiji de lunzheng*, p. 59.

CHAPTER 9

1. Cheng Hong and Jin Yuan, *Sanxia gongcheng da jishi, 1919–1992* (Report on the Three Gorges Project, 1919–1992) (Wuhan: Changjiang wenyi chubanshe, 1992), pp. 160–65.
2. Bureau of Reclamation spokesperson Lisa Guide's telephone interview with Patricia Adams, Probe International, September 23, 1993. http://www.probeinternational.org/pi/documents/three_gorges/who.html
3. Wang Qunsheng, *Sanxia gongcheng fang'an chutai neimu* (Inside Story of Mapping Out the Scheme for the Three Gorges Project) (Chongqing: Chongqing chubanshe, 1992), p. 117.
4. Jiang Di, ed., *Sanxia baiwan yimin chulu hezai: Laizi kuqu de changpian baogao* (What Is the Way Out for the One Million Migrants in the Three Gorges? A Report from the Three Gorges Area) (Chongqing: Chongqing daxue chubanshe, 1992), p. 6.
5. Lu Yuegang, *Changjiang Sanxia: Bange shiji de lunzheng* (The Three Gorges: Debate for Half a Century) (Beijing: Zhongguo shehui kexue chubanshe, 1993), pp. 123,124.
6. Jiang Di, ed., *Sanxia baiwan yimin chulu hezai*, pp. 52, 53, 55.
7. Ibid., p. 58.
8. Jin Xiaoming, *Fengyu Sanxia meng: Sanxia gongcheng jishinian dajishi* (The Three Gorges Dream: A Chronicle of Several Decades of the Three Gorges Project) (Chengdu: Sichuan renmin chubanshe, 1992), p. 25.

9. Shi Wenyin, *Sanxia: Yige kua shiji de meng* (The Three Gorges: A Dream Across the Centuries) (Guangzhou: Huacheng chubanshe, 1992), p. 284.
10. Ibid., p. 287.
11. Ibid., pp. 269–72.
12. Ibid., p. 297.
13. Jin Xiaoming, *Fengyu Sanxia meng*, p. 239.

CHAPTER 10

1. Zhou Weichun, ed., *Meng yuan Yichang* (The Dream Came True in Yichang) (Wuhan: Hubei renmin chubanshe, 1998), p. 45.
2. China Three Gorges Project Construction Yearbook Compilation Committee, *Zhongguo Sanxia jianshe nianjian, 1998* (China Three Gorges Project Construction Yearbook, 1998) (Beijing: Zhongguo Sanxia chubanshe, 1998), p. 3.
3. Yangtze Valley Planning Office, ed., *The Three Gorges of the Yangtze River: A Guide for Tourists* (*Sanxia daguan*) (Beijing: Shuili dianli chubanshe, 1986), p. 337.
4. Richard Hayman, "The New Golden Triangle of China," in Dai Qing, comp., *The River Dragon Has Come!* (New York: M.E. Sharpe, 1998), p. 183.
5. Fred Pearce. "The Biggest Dam in the World," *New Scientist* (London), vol. 145, no. 1962 (January 28, 1995), p. 29.
6. Ibid., p. 25.
7. Yi Si, "The World's Most Catastrophic Dam Failures: The August 1975 Collapse of the Banqiao and Shimanan Dams," in Dai Qing, *The River Dragon Has Come!* p. 33.
8. Ibid., p. 28.
9. "Garbage is Surrounding the Three Gorges Dam Area," *Workers' Daily*, April 16, 2001.
10. Yan Jing, "Preserving the Gorge's Plant Life," *China Daily*, September 20, 1994, p. 11.
11. Erik Eckholm, "Sold China 'Spirit Tree' May Have been Stolen," *The New York Times*, May 16, 1998, sec. A, p. 7.
12. Qing Heng, Li Jian, Hu Tiheng, and Liu Xinyu, "Sanxia kuqu zhian wenti sikao ji duice gouxiang" (Public Security Problems and Countermeasures in the Three Gorges Area) (Internal Report of February 1993 by Officials of Wanxian Prefecture Public Security Department), *Fanzui yu gaizao yanjiu* (Research in Crime and Reform) (Beijing: Fanzui yu gaizao yanjiu bianji bu, 1993), vol. 2, p. 41. The English version of the quotation is from "The Three Gorges Dam in China: Forced Resettlement, Suppression of Dissent and Labor Rights Concerns," *Human Rights Watch/Asia* (New York: Human Rights Watch/Asia), vol. 2, no. 2 (February 1995), p. 25.
13. Ding Qigang, "What Are the Three Gorges Resettlers Thinking?" in Dai Qing, *The River Dragon Has Come!*, p.81.

CHAPTER 11

1. Ding Qigang, "What Are the Three Gorges Resettlers Thinking?" in Dai Qing, comp., *The River Dragon Has Come!* (New York: M.E. Sharpe, 1998), p. 72.
2. Jasper Becker, "Lengthy Sentences Feared for Outspoken Farmers," *South China Morning Post*, April 21, 2001.

CHAPTER 12

1. *Zhongguo Sanxia: The Three Gorges of China* (Beijing: Kexue jishu wenxian chubanshe, 1997), p. 242.
2. Pi Rixiu, "Bianhe huaigu" (Meditating on the Bian River), *Quan tangshi jingpin yizhu huidian* (Anthology of Translations of Selected Tang Dynasty Poems with Notes) (Changchun: Changchun chubanshe, 1994), pp. 916–17. Translated by Shuxi Yin.
3. http://64.124.76.21/BBSView.asp?SubID=mychina&MsgID=25198
4. Jiang Di, ed., *Sanxia baiwan yimin chulu hezai: Laizi kuqu de changpian baogao* (What Is the Way Out for the One Million Migrants in the Three Gorges? A Report from the Three Gorges Area) (Chongqing: Chongqing daxue chubanshe, 1992), pp. 89, 90.
5. Yangtze Valley Planning Office, ed., *The Three Gorges of the Yangtze River: A Guide for Tourists* (*Sanxia daguan*) (Beijing: Shuili dianli chubanshe, 1986), p. 347.
6. *Zhongguo Sanxia: The Three Gorges of China*, p. 256.

APPENDIX I

YANGTZE RIVER FACTS AND FIGURES

THE THIRD-LONGEST RIVER IN THE WORLD after the Nile and the Amazon, and the longest in Asia, the Yangtze flows from west to east for 3,900 miles. It stretches through nine provinces (Qinghai, Tibet, Yunnan, Sichuan, Hubei, Hunan, Jiangxi, Anhui, and Jiangsu) and two municipalities (Chongqing and Shanghai). The Yangtze is usually divided into three parts: the lower reaches, from the mouth to Jiujiang (Hukou) in Jiangxi province (519 miles); the middle reaches, Jiujiang (Hukou) to Yichang in Hubei province (583 miles); and the upper reaches, Yichang to the source in Qinghai province (2,796 miles).

The Three Gorges (Xiling, Wu, and Qutang, from east to west) are located in the upper reaches. Xiling Gorge is in Hubei province, Wu and Qutang Gorges in Sichuan province. In summer, the water level in the gorges is between 500 to 600 feet, making it the one of the deepest stretches of river in the world. Before the rapids in the Three Gorges were blasted away in the 1950s and 1960s, the Yangtze River was generally considered the most treacherous river in the world to navigate and it still ranks among the most difficult. Water levels fluctuate enormously. The water is low in winter, and in summer the melting snows cause a tremendous increase in water level. At Wuhan, there is an average seasonal variation of about 50 feet; at Chongqing, the water can rise by as much as 100 feet in the summer. During China's worst floods, the water level in the gorges has increased by hundreds of feet.

The Yangtze is navigable by large ships between Shanghai and Yibin, a total distance of 1,700 miles. The river is navigable by ships of up to 5,000 tons as far as Wuhan, 3,000 tons to Yichang, 1,500 tons to Chongqing, and 800 tons to Yibin, located 230 miles west of Chongqing. Small boats can continue for about 62 miles beyond that. On the completion of the Three Gorges Dam in 2009, ships of up to 10,000 tons should be able to sail as far as Chongqing.

The Yangtze has a drainage area of an estimated 700,000 square miles, one-fifth of China's total land area, and affects the lives of more than 400 million people. The river carries approximately 35.3 trillion cubic feet of water to the sea annually, and discharges 530 million tons of yellow silt into the Yellow Sea. The river has an estimated hydroelectric potential of 200 million kilowatts.

The Three Gorges Dam will raise the water level by approximately 360 feet to a total depth of 574 feet at the dam site at Sandouping in Xiling Gorge and will create a 360-mile reservoir approximately the same length as Lake Superior. Approximately 1.2 million people will be relocated in the area and to other parts of China.

Appendix II

Three Gorges Dam Facts and Figures

Preparation: 2 years (1992–1994)
Main Construction: 15 years (1994–2009)
Completion: 2009

Length (on top)	1.45 miles
Height	607 feet
Reservoir depth	574 feet
Reservoir length	360 miles
Storage capacity	1.387 trillion cubic feet, of which 782 billion cubic feet is for flood control
Discharging capacity	4.1 million cubic feet per second
Number of generators	26
Total generator capacity	18.2 million kilowatts
Annual output	84.7 billion kilowatt hours (12 times that of Niagara Falls)
Ship lock	Vessels up to 10, 000 tons
Ship Elevator	Vessels under 3, 000 tons
Excavated rock and soil	3.62 billion cubic feet
Fill	1.03 billion cubic feet
Cement used	10.5 billion tons

Timber used	56 million cubic feet
Metal used	1.9 million tons rolled steel
Estimated total cost	24.6 billion U.S. dollars
People displaced	Approximately 1.2 million
Land submerged	243 square miles (13 cities, 140 towns, 1, 351 villages, and 657 factories)
Cultivated land submerged	74,100 acres

APPENDIX III

IMPACT OF THE THREE GORGES DAM ON SELECTED LOCATIONS, IN GEOGRAPHIC ORDER FROM WEST TO EAST

Chongqing: Located at the end point of the new reservoir, the water level in Chongqing harbor will increase by about 50 feet. The urban area of Chongqing will lose little land, but the city faces a potentially enormous increase in the accumulation of silt in the harbor as a result of the Three Gorges Dam. Although scientists disagree about the overall amount of sediment that may collect, rather than being washed out to sea, and the extent to which this can be controlled, the most pessimistic estimate is that Chongqing harbor could be clogged by silt within ten to fifteen years of completion of the dam. The main impact on the city, however, is the influx of investment in support of the Three Gorges Dam and the relocation of tens of thousands of migrants from towns and rural counties in the gorges to new high-rise townships and rural areas to the north and west of the central city of Chongqing. Since 1997, when Chongqing was elevated to the status of a municipality, conditions in Chongqing have improved dramatically, though this has been accompanied by the destruction of much of the old city and the relocation of thousands of urban residents, who have been often unwillingly displaced to the outskirts of the city.

Changshou: The county seat and surrounding rural areas will be minimally affected, with only 7.68 square miles out of a total of 543 square miles flooded. Residents of this area will be moved to a newly expanded city center.

Fuling: The low area of the city, located at the juncture of the Wu River and the Yangtze, will be entirely submerged, requiring the relocation of about 68,000 people. White Crane Ridge, where many poems and records dating back over a thousand years are inscribed on the cliff walls of the river, will be submerged hundreds of feet below water once the reservoir is filled. As this national treasure will be lost, the local government plans to build a museum on the new bank containing reproductions of the carvings. The Wu River valley will be flooded as far as Guizhou province, improving river transportation in remote districts. The excavation of the ancient royal tombs of the Ba Kingdom is underway and should be completed before the waters rise.

Fengdu: Mount Mingshan, the tourist site with its temples and depictions of the horrors of hell, will be inundated, leaving the peak as an island, accessible to visitors by boat. The main industrial town of 50,000, now located on the north bank of the Yangtze, will be entirely flooded. The city will move across the Yangtze to the opposite bank.

Shibao Block: The picturesque old town will be completely flooded and its residents moved to formerly agricultural land just downstream of the town. Once the reservoir is filled, the water will reach to the steps of the red twelve-story pagoda for which the town is famous. To protect the wooden pagoda, a retaining wall will be built around it, turning the site into an artificial island accessible by boat. Surrounding villages will also be flooded, and the peasants relocated to nearby higher areas and more distant sites near the city of Chongqing and in Shandong province. The county seat of Zhongxian will be almost completely submerged.

Wanxian: The water level at Wanxian will rise nearly one hundred feet. Over three-quarters of Wanxian, the largest city in eastern Sichuan, will be inundated by the reservoir. Hundreds of high-rise apartment buildings have been built on what was rich farmland to accommodate the tens of thousands of people who must be relocated both from the lower levels of the city and from other communities along the river. A new downtown above the high-water mark has been renovated and expanded, creating a pleasant modern city filled with shops and restaurants. Hundreds of factories from this major industrial and transportation center will be relocated. The rich agricultural lands of the surrounding counties are a major loss from the dam project, and opposition among farmers has led to police action. A copy of the old clock tower in West Hill Park, the symbol of Wanxian for decades, will be rebuilt and placed on higher ground.

Yunyang: One of the poorest counties in the region, Yunyang will have almost no remaining land that is not too steep to cultivate after the reservoir is filled. The new county seat moved 20 miles upstream to Shuangjiang in 1999, at which time 65,000 people, the majority of the urban population, moved as well. Many rural residents are being relocated locally to outlying areas of the county and to Chongqing, and others to distant sites in Jiangxi, northern Jiangsu, Hubei, and Shanghai. Some 10,000 people have been assigned to northern Jiangsu province, which has resulted in highly publicized protests about conditions there. The Zhang Fei Temple on the southern bank of the Yangtze in Yunyang, reputed to have once held Zhang Fei's severed head in a vat of oil, will be taken apart and rebuilt in Shuangjiang.

Kaixian: A remote interior county, Kaixian lies to the north of Yunyang, east of the urban center of Chongqing and west of Wuxi county on the East and South rivers which converge and flow into the Yangtze. Although Kaixian is more than 50 miles from the Yangtze River, it will lose more land than almost any other county in the region. Located in a fertile and relatively low-lying valley, a large part of its land (ten towns, including the county seat) will be submerged once the Three Gorges project is completed. More than 120,000 of its 1.4 million residents will be relocated, accounting for nearly one-tenth of the total to be resettled by the dam project.

Fengjie: The town will be entirely flooded. The majority of the town's 30,000 urban residents are moving to a new site 4.3 miles upstream. Others will move to a site adjacent to Baidicheng or across the Yangtze to the north bank. The Yidou Gate of the city wall at the entrance to the city will be removed to Baidicheng. The 200-year-old Buddhist temple and Protestant church built in 1902 will be lost but rebuilt in the new town. Rural residents are being relocated locally and on the outskirts of Chongqing, Wanxian, and in Fujian and Zhejiang provinces. The city is attempting to photograph and preserve records and artifacts of the old city as much as possible and plans to build a museum about the old town in the new city.

Baidicheng: The increase in the water level will inundate about three-quarters of the hill on which Baidicheng is located, leaving this historic site from the Han and Three Kingdoms eras as an island overlooking Kuimen and Qutang Gorges. Access to Baidicheng and the new Mammoth Museum at Laoguanmiao, the cliff overlooking Qutang Gorge, will be possible by boat. A suspension bridge built in the mid-1990s connecting Baidicheng and Laoguanmiao will have to be removed because it is below the new water level. The carved trackers' paths along the cliff walls of Qutang Gorge will be submerged.

Wushan: The largest city in the gorges, Wushan has an urban population of close to 70,000, most of whom will moved to the new city that has been built directly above the old. Approximately 15,000 rural residents will also have to be relocated. Some will remain at higher elevations within Wushan county, while substantial numbers will go to Hubei, Anhui, and Guangdong provinces. Wushan lies at the confluence of the Yangtze and Daning rivers and is an important center for regional transportation and tourism to the Little Three Gorges. Tourist authorities expect that the increase in the water level will allow the opening of now inaccessible travel destinations. The water level will rise approximately 164 feet here.

Dachang: Located north of Wushan on the Daning River, past the Little Three Gorges, Dachang has the largest group of surviving Ming and Qing houses anywhere in the Three Gorges region. Although the water level here will rise by only seven feet, the entire town must be moved. Thirty-seven historic houses have been chosen for preservation and will be moved to the new site of the town three miles away, where they will become a tourist site. The remaining houses will be destroyed and replaced by modern structures. Approximately 10,000 people from Dachang have been moved to Guangdong province.

Shennongjia: The highland region north of the Yangtze River on the Sichuan-Hubei border is a national forest and biological preserve, long famous for its rare and unusual plants and animals. Shennongjia is under considerable environmental pressure, as roads are being built into the area and the population is increasing. Environmental preservation areas have been set up for endangered plants and animals at three different elevations.

Badong: The water level will rise approximately 230 feet. An industrial city at the edge of Xiling Gorge in Hubei province, Badong has been relocated upstream to sites on opposite sides of the Yangtze not far from the entrance to Wu Gorge, connected by a 600-foot bridge, the shortest on the Yangtze. A total of about 32,000 people must move, of whom two-thirds are urban residents. Almost all will be relocated in the vicinity of the new city.

Zigui: This town of 30,000 adjacent to Xiling Gorge will be completely flooded. The residents of Zigui, the first in the gorges to be relocated, have moved to Maoping on the south bank of the river near Sandouping. Qu Yuan Memorial Hall, already rebuilt once during construction of the Gezhouba Dam in the 1970s, is also moving to Maoping.

APPENDIX IV

BIOGRAPHIC GLOSSARY OF KEY CONTEMPORARY POLITICAL FIGURES

Chen Yi (1901–1972): Born in Lezhi, Sichuan province, he joined the Chinese Communist Party in France in 1923. As commander-in-chief of the Third Front Army, Chen led Communist forces to victory over the Kuomintang along much of China's east coast in 1949. Chen served as mayor of Shanghai from 1949 to 1954. He became vice premier in 1954 and foreign minister in 1958. Chen was persecuted during the Cultural Revolution and died of cancer in 1972.

Chen Yun (1905–1995): Born in Qingpu (now part of Shanghai), Jiangsu province, Chen joined the Chinese Communist Party in 1925. He served as vice premier from 1949 to 1972, and became vice chairman of the CCP Central Committee in 1956 and one of the top seven leaders. A lukewarm supporter of the Great Leap Forward, he reemerged in 1978 as vice chairman of CCP Central Committee and became an influential senior statesman.

Chiang Kai-shek (1887–1975): Born in Fenghua, Zhejiang province, he headed the Nationalist government from 1928 to 1949. In 1918, Chiang joined Sun Yat-sen's newly founded Kuomintang Party, and in 1925 became head of the Whampoa Military Academy in Guangzhou. In 1926, as commander-in-chief of the KMT Revolutionary Army, Chiang launched a military campaign against the northern warlords to unify China. This campaign, known as the "Northern Expedition," ended in 1928, when his troops took Beijing. After the beginning of war with Japan in 1937, Chiang formed an alliance with the Communists against the Japanese invaders. This truce was short-lived, and following the Japanese surrender in 1945, civil war broke out between the Nationalists and Communists, ending in late 1949 when Chiang Kai-shek and his supporters fled to Taiwan. Always hoping to recapture the mainland, Chiang remained as president of the Republic of China on Taiwan until his death in 1975.

Dai Qing (1941-): Trained as a missile engineer at Harbin Military Engineering Institute, she is the adopted daughter of Ye Jianying, one of China's most senior military leaders. Dai Qing was a journalist at *Guangming Ribao* (Enlightenment Daily) in the early 1980s. In the mid-1980s, Dai Qing collected documents from many scientists and economists opposed to the Three Gorges Dam project. Her first book on the Three Gorges Dam and its impact, *Changjiang! Changjiang!* (Yangtze! Yangtze!), was published in early 1989. After the June fourth crackdown in Tiananmen Square, she was imprisoned for ten months and her book banned for allegedly contributing to "turmoil." Dai published a second book on the Three Gorges, *The River Dragon Has Come*, in 1998.

Deng Xiaoping (1904–1997): Born in Guang'an, Sichuan province, he went to France on a work-study program in 1920 and joined the European branch of the Chinese Communist Party in 1924. Returning to China in 1927, he helped organize Communist military forces in the southwestern province of Guangxi in 1929. During the 1945–49 civil war with the Nationalists, Deng was a member of the Second Field Army. In 1952 he was appointed a vice premier and in 1956 became a member of the CCP Politburo Standing Committee and head of the party Secretariat. He was condemned during the Cultural Revolution for criticizing the personality cult of Mao Zedong and for his "liberal" policies on agriculture and industry, but he made a comeback in 1973 as a vice minister, and in 1975 was reappointed to the Politburo Standing Committee. Though Deng was dropped again in 1976, he reappeared in July 1977, when he assumed all of his previous posts and also became chief of staff of the People's Liberation Army. In 1981, he became chairman of the Central Military Commission. In November 1987, he retired nominally from all posts except the Military Commission, a position that he finally yielded in November 1989. Despite this, Deng remained the paramount leader of China from 1978 until his death in 1997.

Huang Wanli (1911–2001): Born in Chuansha (now part of Shanghai) in Jiangsu province, Huang Wanli graduated from the Department of Civil Engineering of Tangshan Communications University in 1932 and received an M.S. in civil engineering from Cornell University in 1935 and a Ph.D. in the same field from the University of Illinois in 1937. Upon returning to China, Huang became the technical supervisor of the Water Resources Department of the All-China Economic Committee, an engineer in the Water Resources Bureau of the Sichuan province government, and engineer-in-chief and director of the Water Resources Bureau of Gansu province in China's northwest. In 1949 Huang was appointed to professorships at Tangshan Railway College and Tsinghua University, where he conducted research in the field of water resources and hydrology, as well as water resources planning. He was a longtime opponent of the Three Gorges Dam.

Jiang Zemin (1926-): Born in Yangzhou, Jiangsu province, Jiang joined the Chinese Communist Party in 1946. Jiang graduated with a degree in electrical engineering from Jiaotong University in Shanghai in 1947, and then studied economics in the U. S. S. R. On his return to China, he worked his way up through various government ministries in Beijing. In 1985 Jiang became mayor of Shanghai. He replaced Zhao Ziyang as general secretary of the party in June 1989 and when Deng retired in November 1989, he placed Jiang in the powerful post of chairman of the Central Military Commission. In 1993, Jiang was named president of China and was reelected in 1998. Until 2002, he held the three top jobs in the country—president, Communist Party general secretary, and chairman of the party's Central Military Commission.

Li Boning (1914-): Born in Renqiu, Hebei province, Li Boning has been a consistent supporter of the Three Gorges Dam project and an important figure in China's water resources bureaucracies for decades. In the late 1950s Li Boning was the deputy director of the Capital Construction Department of the Ministry of Water Conservancy and Electric Power. In 1978, he became vice minister of water conservancy and electric power, and in 1989, the leading member of the Central Flood Prevention and Control Office. In 1988 he was elected vice chairman of the Economic Committee of the Chinese People's Political Consultative Conference.

Li Peng (1929-): The adopted son of Zhou Enlai, Li Peng was born in Chengdu, Sichuan province, to parents active in the Chinese Communist Party, both of whom were executed during the early 1930s. From 1948 to

1954, Li was trained as an electrical engineer in the Soviet Union, and from 1955 to 1979 worked in China on numerous positions in the power industry. In 1982 he became vice minister of the Ministry of Water Conservancy and Electric Power and a member of the Central Committee at the Twelfth CCP Party Congress. In 1985 he was appointed to the Politburo, and in 1987 to its Standing Committee. Li Peng became premier in 1988. In June 1989 he reportedly transmitted the order issued by Deng Xiaoping for troops to use force against prodemocracy demonstrators. His official term as premier expired in March 1998. In 1998 he was appointed to serve as the chairman of the National People's Congress.

Li Rui (1917-): Born in Pingjiang, Hunan province. Li Rui has been a longtime opponent of the Three Gorges Dam. Li Rui was vice minister of electric power from 1955 to 1958, and then vice minister of the newly reorganized Ministry of Water Conservancy and Electric power in 1958 and 1959. In 1956 Li Rui served on the Yellow River Planning Commission under the State Planning Commission. Serving as A corresponding secretary to Mao Zedong in the 1950s, Li Rui was purged for his support for Peng Dehuai's opposition to the Great Leap Forward. Li was rehabilitated in 1979 and appointed vice minister of the power industry and director of the State Bureau of Computers. From 1982 to 1985 he was a member of the CCP Central Committee, and was appointed in 1985 to the Central Advisory Commission, a largely honorific organization established for semiretired party leaders.

Li Xiannian (1909–1992): Born to poor peasants in Hongan, Hubei province, Li Xiannian trained as a carpenter and then joined the Communists in 1927. Rising to the top of the Chinese Communist Party hierarchy as a military commander, Li Xiannian became a member of the Central Committee in 1945, and after 1949 became party secretary of Hubei province. He served as chairman of the Hubei provincial government from 1950 to 1954 and mayor of Wuhan from 1953 to 1954. From 1954 to 1980 he served as a vice premier and was also minister of Finance until 1975. He was made a member of the CCP Politburo in 1956. He continued to serve on the Politburo throughout the Cultural Revolution (1966–76) and remained a central figure in economic and financial affairs through the 1980s. From 1983–88, he was president of the People's Republic of China.

Lin Biao (1907–1971): Born in Huanggang, Hubei province, he joined the Chinese Communist Party in 1925. Considered a strategic military genius,

Lin rose to a high rank during the Sino-Japanese war and China's civil war. Lin replaced Peng Dehuai as defense minister in 1959, and was designated Mao's successor in 1969. In September 1971, after allegedly plotting a coup against Mao, Lin fled by plane. The plane crashed in Outer Mongolia, en route to the Soviet Union, killing everyone on board.

Lin Yishan (1911-): Born in Wendeng, Shandong province, he became director of the Yangtze River Conservancy Commission in 1949 and a member of the Huai River Control Commission in 1950. In 1958 he was appointed director of the Yangtze Valley Planning Office under the Ministry of Water Conservancy and Electric Power. Lin Yishan was purged during the Cultural Revolution but was returned to power in 1974 when he was put in charge of construction of the Gezhouba Dam. In 1979 he was appointed chairman of the Yangtze River Commission.

Liu Bocheng (1892–1986): Born in Kai county, Sichuan province, Liu Bocheng joined the Chinese Communist Party in 1926. In 1931 he joined the Communist forces in the Jiangxi Soviet area. During the Long March, Liu led his troops through the most difficult terrain in Xinjiang. From 1947 to 1948 Liu's troops operated along the railway between Beijing and Hankou. In April 1949 his troops occupied Nanjing and eventually reached Yunnan, after having pushed through Guizhou and Sichuan. From 1951 to 1958 he was director of the PLA Military Academy. Then, his eyesight seriously deteriorated from a war injury, he went into retirement.

Liu Shaoqi (1898–1969): Born in Ningxiang, Hunan province, he joined the Chinese Communist Party in 1921 in Moscow, studied at Moscow's University of the Toiler of the East from 1921 to 1922, and worked in the underground movement in the late 1930s. Liu urged the party to "study Mao Zedong" in 1942 and became second only to Mao in the CCP from 1943 to the Cultural Revolution in 1966. President of the P.R.C. from 1959 to 1967, he was persecuted during the Cultural Revolution and died in prison in 1969, but was posthumously rehabilitated in 1980.

Luo Ruiqing (1906–1978): Born in Nanchong, Sichuan province, he attended Whampoa Military Academy and joined the Northern Expedition in 1926. During the civil war, he served in various posts in the People's Liberation Army. He was vice premier from 1959 until the Cultural Revolution, when he was purged, and he also served as vice chairman of the National Defense

Council from 1965 to 1966. In 1977 Luo was elected member of the CCP Central Committee.

Mao Zedong (1893–1976): Born in Shaoshan, Hunan province. Mao Zedong came from a wealthy peasant background and in 1921 was one of the twelve founding members of the Chinese Communist Party. In 1928 he became a political commissar of the Red Army and rose to power during the Long March, when the Chinese Communist army retreated to Yan'an in the mountains of Shaanxi province. Mao became chairman of the CCP in 1938, a position he held until his death. After the founding of the People's Republic of China, he served also as president of the P.R.C. from 1949 to 1958, when he gave up this position to Liu Shaoqi. In early 1958 Mao initiated the Great Leap Forward, leading to three years of widespread famine in China. In 1966, he launched the Cultural Revolution. Mao died in September 1976.

Qian Zhengying (1923-): Born in the United States and trained as a civil engineer in China, Qian Zhengying was appointed deputy director of the Water Resources Department in the East China Ministry and Administrative Council in 1950. In 1952 she became vice minister of water conservancy, and in 1957 she was made vice chairman of the Commission for Harnessing the Huai River. In 1958 she became vice minister of the new joint Ministry of Water Conservancy and Electric Power. In 1975 Qian was appointed Minister of Water Conservancy and Electric Power and in 1983 visited the United States as part of an energy delegation. In 1982 she was appointed to the CCP Central Committee and reappointed in 1987. Since the mid-1980s, Qian Zhengying has strongly supported construction of the Three Gorges Dam.

Sun Yat-sen (1866–1925): Known as the Father of the Country in both the P.R.C. and Taiwan, Sun was an influential figure in the overthrow of the Qing dynasty in 1911 and the founder of the Kuomintang. Sun served briefly as provisional president of the new republic but was replaced by General Yuan Shikai in early 1912. Sun's party was thereafter based in Guangdong and had little influence until the 1920s, when it began to receive Soviet aid. In 1924 the Kuomintang entered into an alliance with the recently founded Chinese Communist Party, which at that time was heavily Soviet-influenced.

Wang Renzhong (1917–1992): Born in Jiang, Hebei province, Wang joined the Chinese Communist Party in the Jiangxi soviet area in 1933. He served as first secretary of Wuhan from 1949 to 1954, first secretary of Hubei province from 1954 to 1966, and political commissar of the Wuhan Military Region from 1963 to 1967. An important figure early in the Cultural Revolution, he fell out of power but returned to influence following Mao's death in 1976.

Yang Shangkun (1907–1998): Born in Tongnan, Sichuan province, Yang joined the Chinese Communist Party in 1926 and studied in Moscow. He served as director of the CCP's General Office from 1948 to 1965. Yang was purged in late 1965 but survived the Cultural Revolution to become vice chairman of the Military Affairs Commission and president of the P.R.C. in 1988.

Ye Jianying (1897–1986): Born in Meixian, Guangdong province, he joined the Chinese Communist Party in 1924. He was the party secretary of Guangzhou municipality from 1950 to 1954, a Politburo member in 1966, and a Politburo Standing Committee member from 1973 to 1985. In 1975, he was appointed minister of national defense until 1978. Ye was a leading figure in the arrest of the Gang of Four. In 1978 he was elected chairman of the Standing Committee of the National People's Congress.

Yuan Shikai (1859–1916): Born in Xiangcheng, Henan province, Yuan began his career in the army. In 1901 he took charge of the region around Peking. After the Qing dynasty collapsed in 1911, with support from both the emperor in Peking and the provisional president Sun Yat-sen in Nanking, Yuan became the provisional president of China in 1912. Thereafter, he contrived to make himself president for life and then tried to create a new imperial dynasty in late 1915. Widespread opposition challenged his legitimacy, and he died in 1916, shortly after giving up his ambition to reign as emperor.

Zeng Siyu (1911-): Born in Jiangxi province, he joined the Chinese Communist Party in 1932 and participated in the Long March. He rose to high rank in the Sino-Japanese, civil, and Korean wars. During the 1950s he served as vice commander of the Shenyang Military Region. In 1967 he was appointed commander of the Wuhan Military Region; in 1968, with the establishment of the Revolutionary Committee of Hubei province, Zeng became chairman; in 1971, he was chosen as first party secretary of Hubei province, serving in all three posts until 1973. In 1974 he was appointed commander of the Ji'nan Military Region and remained in this position until his retirement in 1980.

Zhang Guangdou (1912-): Born in Changshu, Jiansu province, Zhang graduated from the Civil Engineering College of Shanghai Jiaotong University in

1934. He received an M.S. from the University of California in 1936 and another M.S. from Harvard University in 1937. Since the 1940s, Zhang has been engaged in teaching and research on water conservancy and hydroelectric engineering. Zhang was the chief designer of many of China's most important hydropower projects on the Yellow and Yangtze rivers. These included the Gezhouba, Danjiangkou, Sanmenxia, and Xiaolangdi dams and the Three Gorges project. He is also a professor at Tsinghua University in Beijing.

Zhou Enlai (1898–1976): Born in a well-to-do gentry family in Huaian, Jiangsu province, Zhou served as foreign minister and premier after 1949. He endorsed a liberalization of policies toward intellectuals in the 1950s. Zhou was loyal to Mao Zedong throughout his career and during the Cultural Revolution reluctantly supported widespread purges of party leaders but attempted to protect many people from Red Guard attacks. Zhou also provided protection for China's historical relics, some along the Yangtze, which were often targeted for destruction by rampaging Red Guards.

Zhu De (1886–1976): Born in Yilong, Sichuan province, he joined the Chinese Communist Party in Germany in 1922. In 1930 Zhu was elected chairman of the Military Council of the CCP Central Committee. Zhu was commander-in-chief of the Red Army in the 1930s and commander-in-chief of the People's Liberation Army after the founding of the P.R.C. until 1954. In 1949 he was appointed vice chairman of the Government Council. In 1954 he was elected vice chairman of the P.R.C., a post he held until 1959. He was a Politburo Standing Committee member from 1956 to 1976.

Zhu Rongji (1928-): Born in Changsha, Hunan province, Zhu joined the Chinese Communist Party in 1949. In 1951, he graduated from Tsinghua University, where he studied electrical engineering. Zhu was labeled a rightist in 1958. Rehabilitated in 1962, he worked as an engineer until 1969. In 1970, during the Cultural Revolution, Zhu was purged again and underwent "reeducation" until 1975. From 1975 to 1978, Zhu served as an engineer in the Electric Power Communications Company and from 1978 to 1979 as director of the Industrial Economics Institute under the Chinese Academy of Social Sciences. Zhu worked for the State Economic Commission (SEC) and the Comprehensive Bureau from 1979 to 1982, he then became director of the SEC's Bureau of Technical Information. In 1983 he was promoted vice minister of the SEC, where he remained until 1987. From 1987 to 1991, he served as deputy secretary, then secretary of the CCP Shanghai Municipal

Committee. In 1991 Zhu became vice premier of the State Council. In 1998 Zhu became premier.

Zou Jiahua (1926-): Born in Shanghai, Zou joined the Chinese Communist Party in 1945 and studied machinery manufacturing in Moscow. Zou Jiahua served in the 1950s and 1960s as a director of a machine tool plant in northeast China and then worked in the First Ministry of Machine Building. In 1977 he became a deputy director of the National Defense Office under the State Council. In 1982 he became vice minister of the Commission of Science, Technology and Industry for National Defense, and was appointed minister of the ordnance industry. In 1988 he became a state councilor and minister of machine building and the electronics industry. In 1991 he became head of the State Planning Commission and a vice minister and in 1992 a member of the Politburo at the Fourteenth Party Congress. Since March 1998 he has served as the vice chairman of the standing committee of the National People's Congress.

BIBLIOGRAPHY

BOOKS IN ENGLISH

Alley, Rewi. *Man Against Flood.* Beijing: New World Press, 1956.

Banister, Judith. *China's Changing Population.* Stanford: Stanford University Press, 1987.

Becker, Jasper. *Hungry Ghosts: Mao's Secret Famine.* New York: Free Press, 1996.

Bernstein, Thomas P. *Up to the Mountains and Down to the Villages: The Transfer of Youth from Urban to Rural China.* New Haven: Yale University Press, 1977.

Bird, Isabella. *The Yangtze Valley and Beyond: An Account of Journeys in China, Chiefly in the Province of Sze Chuan and Among the Man-Tze of the Somo Territory.* London: John Murray, 1899.

Blakiston, Thomas W. *Five Months on the Yang-Tsze, and Notices of the Present Rebellions in China.* London: John Murray, 1862.

Bonavia, David. *China's Warlords.* Hong Kong: Oxford University Press, 1995.

Bonavia, Judy. *The Yangzi River.* Rev. by Madeleine Lynn. Hong Kong: The Guidebook Company, 1995.

Changjiang Branch of China International Travel Service. *Along the Changjiang River.* Wuhan: 1982.

China Tames Her Rivers. Beijing: Foreign Languages Press, 1972.

China Yangtze Three Gorges Project Development Corporation in Yichang. *Three Gorges Projects.* 2000.

Cressey, George Babcock. *China's Geographic Foundations: A Survey of the Land and its People.* New York: McGraw-Hill, 1934.

Dai, Qing, comp. *The River Dragon Has Come! The Three Gorges Dam and the Fate of China's Yangtze River and Its People.* Ed. John G. Thibodeau and Philip B. Williams. New York: M.E. Sharpe, 1998.

———,ed. *Yangtze! Yangtze!* Ed. Patricia Adams and John Thibodeau. London and Toronto: Earthscan, 1994. (Translation of 1989 Chinese edition.)

Davidson, Robert J. and Isaac Mason. *Life in West China: Described by Two Residents in the Province of Sz-Chwan.* London: Headley Bros., 1905.

Duncan, Marion H. *The Yangtze and the Yak.* Alexandria, VA: Edward Bros., 1952.

Fitkin, Gretchen Mae. *The Great River: The Story of a Voyage on the Yangtze Kiang.* Shanghai: North China Daily News & Herald and Kelly & Walsh, 1922.

Gill, William. *The River of Golden Sand: The Narrative of a Journey through China and Eastern Tibet to Burmah.* Vol. 1. London: John Murray, 1880.

Goodman, David S. G., ed. *China's Provinces in Reform: Class, Community, and Political Culture*. London: Routledge, 1997.

Graham, A. C., trans. *Poems of the Late T'ang*. Baltimore, MD: Penguin Books, 1965.

Haines, Gregory. *Gunboats on the Great River*. London: Macdonald and Jane's, 1976.

Hersey, John. *A Single Pebble*. New York: Vintage Books, 1989.

Hewlett, Meyrick. *Forty Years in China*. London: Macmillan, 1943.

Hosie, Alexander. *Three Years in Western China: A Narrative of Three Journeys in Ssu-ch'uan, Kuei-chow, and Yun-nan*. London: George Philip & Son, 1897.

Jing, Jun. *Three Gorges Project Briefing Package*. Berkeley, CA: International Rivers Network, 1998.

———. *The Temple of Memories: History, Power, and Morality in a Chinese Village*.Stanford, CA: Stanford University Press, 1996.

Kane, Penny. *Famine in China, 1959–61: Demographic and Social Implications*. New York: St. Martin's Press, 1988.

Kapp, Robert A. *Szechwan and the Chinese Republic: Provincial Militarism and Central Power, 1911–1938*. New Haven, CT: Yale University Press, 1973.

Li, Zhisui. *The Private Life of Chairman Mao: The Memoirs of Mao's Personal Physician*. Trans. Tai Hung-chao. Editorial assistance Anne F. Thurston. New York: Random House, 1994.

Lieberthal, Kenneth and Michel Oksenberg. *Policy Making in China: Leaders, Structures, and Processes*. Princeton, NJ: Princeton University Press, 1988.

Little, Archibald John. *Through the Yang-tse Gorges, or, Trade and Travel in Western China*. 3rd rev.ed. London: Sampson Low, Marston, Serle, & Rivington, Ltd., 1898.

Liu, Zuozhong. *Thousands of Miles: Three Gorges*. Chengdu: Sichuan meishu chubanshe.

Lo, Kuan-chung. *Three Kingdoms: China's Epic Drama*. Trans. Robert Moss. New York: Pantheon, 1976.

Lodwick, Kathleen L. *Crusaders Against Opium: Protestant Missionaries in China, 1874 - 1917*. Lexington: University Press of Kentucky, 1996.

Lynn, Madeleine, ed. *Yangtze River: The Wildest, Wickedest River on Earth: An Anthology*. Hong Kong: Oxford University Press, 1997.

Ma, Ning. *Our Great Three Gorges Project (Da Sanxia xieyi)*. Hong Kong: Hong Kong World Chinese Arts Publication, 2001.

Margary, A. R. *The Journey of Augustus Raymond Margary: From Shanghae to Bhamo, and Back to Manwyne*. London: Macmillan, 1876.

Oliphant, Laurence. *Elgin's Mission to China and Japan*. Vol. 2. 1859. Reprint, Hong Kong: Oxford University Press, 1970.

Peck, Graham. *Two Kinds of Time*. Boston: Houghton Mifflin, 1950.

———. *Through China's Wall*. Boston: Houghton Mifflin, 1940.

Plant, Cornell. *Glimpses of the Yangtze Gorges*. Shanghai: Kelly & Walsh, 1921.

———. *Handbook for the Guidance of Shipmasters on the I-chang-Chungking Section of the Yangtze River*. Shanghai: Maritime Customs, 1920.

Ryder, Granne, ed. *Damming the Three Gorges: What Dam-Builders Don't Want You To Know: A Critique of the Three Gorges Water Control Project Feasibility Study*. London and Toronto: Earthscan Publishers, 1990.

Sage, Steven F. *Ancient Sichuan and the Unification of China*. Albany: State University of New York Press, 1992.

Schneider, Laurence A. *A Madman of Ch'u: The Chinese Myth of Loyalty and Dissent*. Berkeley: University of California Press, 1980.

Shapiro, Judith. *Mao's War Against Nature: Politics and the Environment in Revolutionary China*. New York: Cambridge University Press, 2001.

Shen, Congwen. *Recollections of West Hunan*. Trans. Gladys Yang. Beijing: Panda Books, 1982.

Spence, Jonathan. *The Search for Modern China*. New York: Norton, 1990.

Tolley, Kemp. *Yangtze Patrol*. Annapolis, MD: Naval Institute Press, 1971.

Tretiakov, Sergei M. *A Chinese Testament: The Autobiography of Tan Shih-hua, as Told to S. Tretiakov*. Reprint, Westport, CT: Hyperion Press, 1978.

Van Slyke, Lyman P. *Yangtze: Nature, History, and the River*. Reading, MA: Addison-Wesley, 1988.

von Glahn, Richard. *The Country of Streams and Grottoes: Expansion, Settlement, and the Civilizing of the Sichuan Frontier in Song Times*. Cambridge: Council on East Asian Studies, Harvard University, 1987.

Waley, Arthur. *One Hundred & Seventy Chinese Poems*. New York: Alfred A. Knopf, 1923.

Walters, Derek. *Chinese Mythology: An Encyclopedia of Myth and Legend*. London: Diamond Books, 1995.

Wang, Hongying. "Comparative Studies of Flood Control Measures in China and in the United States." M. A. thesis, York University, 1993.

White, Theodore H. and Annalee Jacoby. *Thunder Out of China*. New York: William Sloane, 1946.

Winchester, Simon. *The River at the Center of the World: A Journey up the Yangtze and Back in Chinese Time*. New York: Henry Holt, 1996.

Wong, How Man. *Exploring the Yangtse*. Hong Kong: Odyssey Productions, 1989.

Woodhead, H.G.W. *The Yangtsze and Its Problems*. Shanghai: Mercury Press, 1931.

Woodrooffe, Thomas. *The River of Golden Sand*. London: Faber and Faber, 1936.

Worcester, G. R. G. *The Junks and Sampans of the Yangtze*. Reprint, Annapolis, MD: Naval Institute Press, 1971.

———. *Sail and Sweep in China*. London: H.M.S.O., 1966.

Wu, Cheng'en. *Journey to the West*. Trans. W. J. F. Jenner. Beijing: Foreign Languages Press, 1986.

Yang, Xianyi and Gladys Yang, trans. *Poetry and Prose of the Tang and Song*. Beijing: Panda Books, 1984.

Yin, Liangwu. "The Long Quest of Greatness: China's Decision to Launch the Three Gorges Project." Ph.D. diss., Washington University in St. Louis, 1996.

Zhang, Xinxin and Ye Sang, eds. *Chinese Profiles*. Trans. Gladys Yang. Beijing: Panda Books, 1986.

Articles and Book Chapters in English

Adams, Patricia. "Planning for Disaster: China's Three Gorges Dam." *Multinational Monitor*, September 1994.

Becker, Jasper. "Whip Comes Down on Yangtze Kickback Officials." *South China Morning Post* (Hong Kong), February 14, 1999.

Brauchli, Marcus W. "River of Dreams: The Mighty Yangtze Seizes a Major Role in China's Economy." *Wall Street Journal*, December 13, 1995.

Chang, Chih-ching. "A Hundred Schools Contend." *China Reconstructs*, vol. 6, no. 1 (January 1957).

Chang, Yen-wei. "Through the Yangtze Gorges." *China Reconstructs*, vol. 5, no.1 (January 1956).

Chao, Liang. "Blocking Yangtze: 'No Problem'." *China Daily*, November 8, 1997.

Childs-Johnson, Elizabeth, Joan Lebold Cohen, and Lawrence R. Sullivan. "Race Against Time: Chinese Scholars Scramble to Save Sites Threatened by World's Biggest Dam." *Archaeology*, November/December 1996.

Chu, Coching. "What China's Scientists Are Doing." *China Reconstructs*, vol. 5, no. 3 (March 1956).

Ciochon, Russell L. "The Ape That Was: Asian Fossils Reveal Humanity's Giant Cousin." *Natural History*, November 1991.

Fisher, Arthur. "World's Largest Dam." *Popular Science*, August 1996.

Goldstein, Carl. "Gorged on Power: Yangtze Dam Boosts Attraction of Dongfang Electrical." *Far Eastern Economic Review*, March 24, 1994.

Guo, Anding. "Old and New in Yichang and Xiangfan." *China Reconstructs*, vol. 38, no. 11 (November 1989).

Harding, James. "Three Gorges Seeks to Open Financing Floodgates." *Financial Times*, August 6, 1998.

Hayman, Richard. "Life and Death of the Yangzi River." Unpublished manuscript for the Association for Asian Studies Annual Conference, East-West Center, University of Hawaii. April 1996.

Heath, Ray and Foo Choy Peng. "An Economy That Won't Add Up." *South China Morning Post* (Hong Kong), August 6, 1994.

Hoh, Erling. "The Long River's Journey Ends." *Natural History*, July 1996.

Hsueh, Pao-ting. "We Are Building Better and More Quickly." *China Reconstructs*, vol. 5, no. 6 (June 1956).

Hu, Tan-fu. "First Bridge Across the Yangtze." *China Reconstructs*, vol. 6, no. 10 (October 1957).

Hua, Joan. "Harnessing the Yangtze." *Window*, November 5, 1993.

Human Rights Watch / Asia. *The Three Gorges Dam in China: Forced Resettlement, Suppression of Dissent and Labor Rights Concerns.* New York: Human Rights Watch, February 1995.

Huus, Karl. "More Dam Trouble." *Far Eastern Economic Review*, October 20, 1994.

Jing, Jun. "Displacement, Resettlement, Rehabilitation, Preparation and Development. China Report" (Draft). Working paper prepared for the World Commission on Dams. August 1999.

Kosowatz, John J. "Mighty Monolith." *Extreme Engineering*, February 2000.

Kuhn, Anthony, and Lincoln Kaye. "Bursting at the Seams: Rural Migrants Flout Urban Registration System." *Far Eastern Economic Review*, March 10, 1994.

Li, Jung. "An Account of the Salt Industry in Tzu-liu-ching." Trans. Fang Tu Lienche. *Science and Technology in East Asia*, ed. Nathan Sivin. New York: Science History Publications, 1977.

Lin, Yishan. "Thirty Years of Controlling China's Greatest River." *China Reconstructs*, vol. 28, no. 10 (October 1979).

Meng, Leng. "The Battle of Sanmenxia: Population Relocation During the Three Gate Gorge Hydropower Project." *Chinese Sociology and Anthropology*, ed. Jun Jing, vol. 31, no. 3 (Spring 1999).

Meng, Xiaoshu. "Report Alleges Forced Labour Is Being Used in Construction: Dam Investigators Warned Off by Human Rights Group." *Hong Kong Standard*, February 22, 1995.

Needham, Joseph. "Civil Engineering and Nautics." In *Science and Civilization in China*. vol. 4, pt. 3. New York: Cambridge University Press, 1971.

Pearce, Fred. "The Biggest Dam in the World." *New Scientist* (London), vol. 145, no. 1962 (January 28, 1995).

Plafker, Ted. "Gorges Disaster Warning." *South China Morning Post* (Hong Kong), February 22, 1995.

Service, Caroline. "Two Trips Through the Yangtze Gorges." *China Reconstructs*, vol. 27, no. 9 (September 1978).

Spence, Jonathan. "A Flood of Troubles." *The New York Times Magazine*, January 5, 1997.

Staff reporter. "The Hanchiang River Story." *China Reconstructs*, vol. 23, no. 6 (June 1974).

Sullivan, Lawrence R. "The Three Gorges Dam and the Issue of Sustainable Development in China." In *The Global Environment: Institutions, Law, and Policy*. Ed. Norman J. Vig and Regina S. Axelrod (Washington, DC: CQ Press, 1999), pp. 300–16.

———. "The Three Gorges Project: Dammed if They Do?" *Current History*, September 1995.

Topping, Audrey R. "Ecological Roulette: Damming the Yangtze." *Foreign Affairs*, vol. 74, no. 5 (September-October 1995).

Tung, Shao-sheng. "The Yangtze Belongs to the People." *China Reconstructs*, vol. 9, no. 3 (March 1960).

Tyler, Patrick. "China's Endless Task to Stem Centuries of Flood." *The New York Times*, September 15, 1996.

———. "Tapping China's Troubled Waters." *The New York Times*, May 24, 1996.

———. "Cracks Show Early in China's Big Dam Project." *The New York Times*, January 15, 1996.

———. "Countdown for Trips Along the Yangtze River." *The New York Times*, March 5, 1995.

Walker, Tony. "Power Hungry on the Gorge." *Financial Times*, February 17, 1997.

Wu, Ming. "Disaster in the Making?" *China Rights Forum*, Spring 1998.

Xie, Jian. "China: Thousands of Dams Reportedly Need Urgent Repairs." *Greenwire*, March 26, 1999.

Yan, Jing. "Preserving the Gorge's Plant Life." *China Daily*, September 20, 1994.

Zich, Arthur. "China's Three Gorges Before the Flood." *National Geographic*, September 1997.

BOOKS AND MONOGRAPHS IN CHINESE

An Ran, ed. *Changjiang Sanxia gongcheng baiwen* (One Hundred Questions About the Yangtze River's Three Gorges Project). Beijing: Guangming ribao chubanshe.

Cao Yingwang. *Zhou Enlai yu zhishui* (Zhou Enlai and Water Conservancy). Beijing: Zhongyang wenxian chubanshe, 1991.

China Three Gorges Project Corporation. *Jianshe zhong de Sanxia* (Three Gorges Project under Construction). Wuhan: Changjiang nianjian chubanshe, 2001.

Committee on Fengjie County Records Compilation. *Fengjiexian zhi* (Fengjie County Records). Beijing: Fangzhi chubanshe, 1995.

Committee on Wanxian County Records Compilation. *Wanxian zhi* (Wanxian County Records). Chengdu: Sichuan cishu chubanshe, 1995.

Committee on Wushan County Records Compilation. *Wushanxian zhi* (Wushan County Records). Chengdu: Sichuan renmin chubanshe, 1991.

Committee on Yangtze River Yearbook Compilation. *Changjiang nianjian* (Yangtze River Yearbook), 1994–1995, Volumes 1 and 2. Wuhan: Changjiang nianjian chubanshe, 1996.

Committee on Zhong County Records Compilation. *Zhongxian zhi* (Zhong County Records). Chengdu: Sichuan cishu chubanshe, 1994.

Gong Li. *Ninghe fengqing* (Ning River Scenery). Wushan: unpublished manuscript, 1994.

Hu Hongbao. *Guanyu Sanxia wenwu baohu ruhe shiying shehui zhuyi shichang jingji de sikao* (Thoughts about How Three Gorges' Antiques Protection Adapts to the Socialist Market Economy). In *Zhongguo Sanxia jianshe* (China Three Gorges Construction). Yichang: Zhongguo Sanxia jianshe zazhi chubanshe, 2001.

Huang Zhigang. *Zhongguo de shuishen* (China's Water Gods). Hong Kong: Longmen chubanshe, 1968.

Jiang Di, ed. *Sanxia baiwan yimin chulu hezai: Laizi kuqu de changpian baogao* (What Is the Way Out for the One Million Migrants in the Three Gorges? A Report from the Three Gorges Area). Chongqing: Chongqing daxue chubanshe, 1992.

Jin Xiaoming. *Fengyu Sanxia meng: Sanxia gongcheng jishinian dajishi* (The Three Gorges Dream: A Chronicle of Several Decades of the Three Gorges Project). Chengdu: Sichuan renmin chubanshe, 1992.

Lai Yong. *Changjiang lüyou* (Yangtze River Tourism). Chengdu: Chengdu ditu chubanshe, 1999.

Li Hanjie, ed. *Zhongguo fen shengshixian dacidian* (Grand Dictionary for China's Provinces, Cities, and Counties). Beijing: Zhongguo lüyou chubanshe, 1990.

Liu Dingxiang, ed. *Zhongguo chengshi baike quanshu—Yichang shi* (Encyclopedia of Chinese Cities—Yichang City). Beijing: Guangming ribao chubanshe, 1988.

Lu Yuegang. *Changjiang Sanxia: Bange shiji de lunzheng* (The Three Gorges: Debate for Half a Century). Beijing: Zhongguo shehui kexue chubanshe, 1993.

Luo Chengzheng and Yue Jiaxiang, eds. *Zhongguo dahongshui: Zaihaixing hongshui shuyao* (Floods of China: A Summary of Catastrophic Floods). Beijing: Zhongguo shudian, 1996.

Ma Shutian. *Chaofan shijie: Zhongguo simiao erbai shen* (Unconventional World: 200 Gods in Chinese Temples). Beijing: Zhongguo wenshi chubanshe, 1990.

Ma Youming. *Shibaozhai de chuanshuo* (Tales of Shibaozhai). Chengdu: Sichuan wenyi chubanshe, 1986.

Office of Fengjie County Records. *Fengjie nianjian, 1991–1996* (Fengjie Yearbook, 1991–1996). Chengdu: Bashu chubanshe, 1998.

Shi Mingding and Luan Linbin, eds. *Changjiang* (The Yangtze River). Shanghai: Shanghai jiaoyu chubanshe, 1989.

Shi Wenyin. *Sanxia: Yige kua shiji de meng* (The Three Gorges: A Dream Across the Centuries). Guangzhou: Huacheng chubanshe, 1992.

Song Huajiu. *Zoujin Sanxia* (Entering the Three Gorges). Wuhan: Wuhan daxue chubanshe, 1995.

Su Chuanqu. *Manhua Fengjie* (About Fengjie). Beijing: Sanxia chubanshe, 1995.

Wang Jie et al., eds. *Changjiang dacidian* (Grand Dictionary of the Yangtze River). Wuhan: Wuhan chubanshe, 1997.

Wu Shuye. *Wushan lansheng* (Wonderful Scenery of Mount Wushan). Chongqing: Chongqing chubanshe, 1988.

Yan Gang and Li Shikui, eds. *Guicheng shilian xuanji* (Selected Poetry of the Ghost City). Fengdu: Fengdu Mingshan guanli weiyuan hui.

Yangtze River Water Resource Association, ed. *Sanxia gongcheng de lunzheng* (Three Gorges' Feasibility Investigation). Beijing: Zhongguo shehui kexue chubanshe, 1990.

Yao Qiuyun. *Fengdu qiwen* (Strange Tales of the Ghost City). Chengdu: Tiandi chubanshe, 1996.

Yuan Gang, ed. *Guicheng Fengdu: Zhongguo shenqu zhi xiang* (Fengdu, The City of Ghosts: Home of the Chinese Divine Comedy). Chengdu: Sichuan yishu chubanshe, 1995.

Yuan Ran. *Nüwa, laozhai, fengqingjie* (Short Fiction from the Three Gorges Area). Hong Kong: Wenguang chubanshe, 1993.

Zeng Xiuqiao and Yang Dekun. *Fengjiexian zhi* (Fengjie County Records). Taipei: Taiwan xuesheng chubanshe, 1971.

Zhang Wenyu. *Baidi cheng* (The White Emperor Town). Chengdu: Sichuan renmin chubanshe, 1995.

Zhang Xinglian, comp. *Zhongguo kaogu cihui* (A Glossary of Chinese Archaeology). Beijing: Waiwen chubanshe, 1983.

Zhou Weichun, ed. *Meng yuan Yichang* (The Dream Came True in Yichang). Wuhan: Hubei renmin chubanshe, 1998.

Zhu Nong, ed. *Sanxia gongcheng yimin yu kuqu fazhan yanjiu* (Research on the Three Gorges Project: Resettlement and Development of the Reservoir Area). Wuhan: Wuhan daxue chubanshe, 1996.

Zhu Nong and Wang Bing, eds. *Changjiang diqu renkou wenti yu kechixu fazhan* (Changjiang River Regional Population Problems and Sustainable Development). Wuhan: Wuhan chubanshe, 1999.

INDEX